AF180676

Aston Postural Assessment Workbook

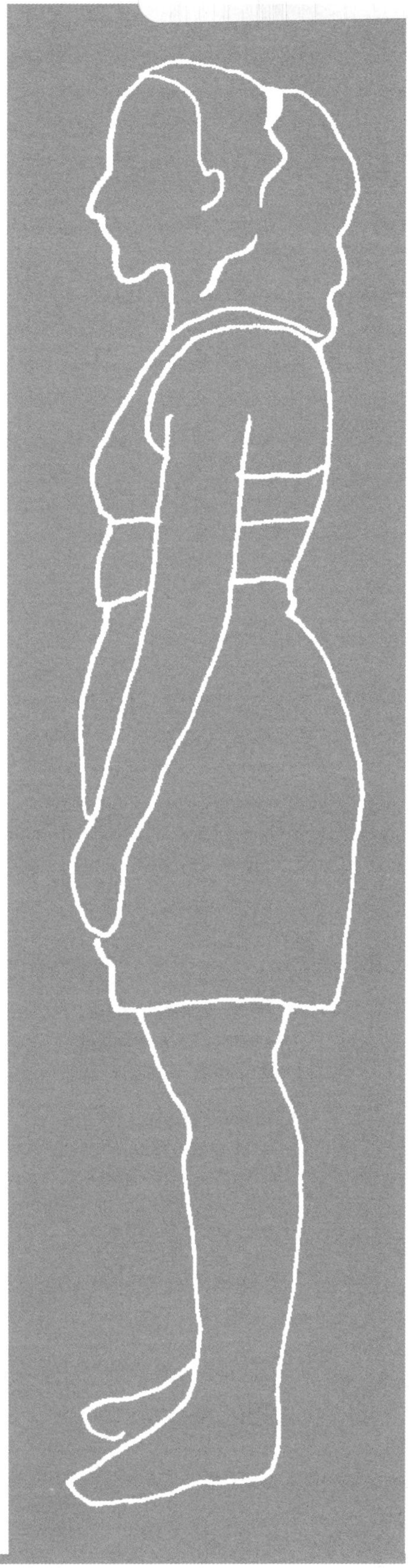

Skills for Observing and Evaluating Body Patterns

Judith Aston, M.F.A.

Illustrations by
Elizabeth Webber

Therapy
Skill Builders®
a division of
The Psychological Corporation
555 Academic Court
San Antonio, Texas 78204-2498
1-800-228-0752

Reproducing Pages from This Book

As described below, some of the pages in this book may be reproduced for instructional use (not for resale). To protect your book, make a photocopy of each reproducible page. Then use that copy as a master for photocopying.

Copyright © 1998 by
Therapy Skill Builders®
a division of
The Psychological Corporation
555 Academic Court
San Antonio, Texas 78204-2498
1-800-228-0752

All rights reserved. No part of this publication may be reproduced or transmitted in any form or by any means, electronic or mechanical, including photocopy, recording, or any information storage and retrieval system, without permission in writing from the publisher.

Permission is hereby granted to reproduce the Observation Sequence, the Sample Session Sequence, and the Sample Session Form in this publication in complete pages, with the copyright notice, for administrative and/or instructional use and not for resale.

The Learning Curve Design and *Therapy Skill Builders* are registered trademarks of The Psychological Corporation.

Printed in the United States of America.

076161530X

2 3 4 5 6 7 8 9 10 11 12 A B C D E

Visit our Web site at www.hbtpc.com.

This book is dedicated
to my mother,

Edith Opal White

About the Author

The genesis of Judith Aston's ideas about what would later become Aston-Patterning® came while she was teaching movement classes at a community college in 1963. There she taught students awareness of their posture and movement patterns. She earned a B.A. and an M.F.A. as well as a Lifetime Teaching Credential in Dance and Physical Education from the University of California at Los Angeles. She also is trained in massage therapy.

In 1968, Ms. Aston's experience in teaching and designing movement programs caught the attention of the renowned Dr. Ida Rolf—founder and developer of the connective-tissue system of body rehabilitation known as Rolfing®. Dr. Rolf asked her to design a movement component to Rolfing, and Ms. Aston designed and taught *seeing* and movement analysis to Rolfing students from 1971–1977.

Today, in addition to lecturing and conducting workshops on Aston-Patterning in the United States, Europe, and Asia, Ms. Aston designs ergonomic products for work, sports, and the home, and is listed in the official Who's Who of American Inventors. She also is a member of the American Massage Therapy Association, the International Dance and Exercise Association, and the Associated Bodywork and Massage Professionals.

Ms. Aston trains others to teach her principles, and she conducts classes on her system of wellness programs in Lake Tahoe, Nevada.

Table of Contents

Foreword

In these analyses of postural alignment and muscle balance, Judith Aston presents the first part of her published work on Aston-Patterning, which is her own creation. This excellent book, originally a synthesis of the author's unique teaching courses, has now matured into a very practical text, a triumph of orderly thought and presentation. This study presents only one aspect of Aston-Patterning, which is an integrated system of movement education, three-dimensional soft tissue work, fitness training, and ergonomics. It is unique among the somatic disciplines in its completeness. Aston-Patterning includes a comprehensive formal evaluative system, innovative manual techniques that are applicable to a wide range of dysfunction in the musculoskeletal system, a number of exercise programs designed to develop a particular aspect of fitness, movement education procedures, and an array of fundamental concepts that allow the work to have an extraordinary scope of application.

Aston-Patterning presents a system of therapy that has grown out of creative observation and consistent, independent development. One that, in its manner of observing and interpreting human movement and alignment, I have long regarded as a fundamental contribution to the field of a variety of therapies. Although no explicit references are cited, this work accords with the findings of modern neurophysiology and biomechanics in every field. The preventive and therapeutic value of economical posture and movement, the conception of the human being as an organism constantly reacting with gravity, ground reaction force, and other external stimuli are inherent in this system.

The major focus of this text is viewing static posture and alignment containing a series of exercises for learning, in order for the therapist to facilitate a greater appreciation of the structure of the human body. The thoroughness of presentation gives the therapist immediate access to meaningful, practical work and will allow him or her to form a basis from which to extend his or her knowledge of human structure to economical movement therapy—*seeing* structurally, from the aspect of alignment, dimension, and proportion. Alignment deals with the placement of the body and its parts. The relationship of parts in space and the distribution of weight bearing through these parts are two aspects of alignment. Dimension and proportion concern the internal volume of the body or a body part, and how that volume is distributed with regard to its component dimensions of width, length, and depth. The three-dimensional position in space of each skeletal part can also be determined. Deviations from normal are expressed in biomechanical terms such as rotation, shear, and compression.

Another component of Aston-Patterning is the correlation of alignment patterns to muscle tone. In assessing muscle tension, the examiner observes and palpates the tension patterns throughout the body and notates areas where there is excessive or lack of tension that might be inappropriate. This body map creates a visual record of the body's patterns.

The level of integration between body work, movement education, and ergonomics, as well as the recognition of asymmetry in structure and movement, and the individual problem solving for each patient are unique to this approach. This text introduces Aston *seeing* skills as a basis for evaluating posture. The thoroughness in presentation of this work gives the therapist immediate access to meaningful, practical work and will enable him or her to develop an unlimited and eminently individualized program of patient care.

It should be remembered that what we see is profoundly influenced by the way we look. Every individual tends to take the limits of his or her own perception also to be the limits of the perceived world, and a cardinal error is to see clinical problems only as one's own idiom. A healthy eclectic approach offsets this otherwise limited trend.

–Darlene Hertling

Preface

Soon after I began teaching movement at a community college in 1963, I was asked to teach stage movement for the theater department. It quickly became apparent that my young acting students were limited in their character expression by their lack of awareness about their own body posture and movement patterns. They needed skills to see posture and movement in order to build a different character expression from their own habitual expression.

A car accident in 1966 from which I suffered serious injuries led to my association with Dr. Ida Rolf and her innovative connective-tissue system of body rehabilitation known as Rolfing. When I had my first Rolfing session with her in 1968, I knew immediately that I had found help toward physical rehabilitation. During that session, Dr. Rolf learned of my background in designing movement programs, and she asked me to develop a movement-education component for Rolfing. In the summer of that year, I began training classes with Dr. Rolf. In this first class, the students would ask me to explain how I could *see* body patterns, and their interest grew. In 1971, I created a class and began to teach *seeing* and movement analysis to all Rolfing students as part of their training.

It is now 1998 and although my idea about a biomechanical model for the body's structure and function has changed significantly, my step-by-step process of teaching *seeing* has proven helpful to many. I hope this also will be true for you.

Acknowledgments

This first book has been 30-plus years in the making and, at times, I wondered if it would take another 30 years. I wish to thank the friends, colleagues, students, and teachers who generously assisted me in this project. The following people fall under all of the above categories: Linda Krier, Allison Funk, Joani Gelinas, and Ronnie Oliver, for ideas in clarifying, editing, and overall support. As well, I want to thank Michael Galbraith for his assistance with layout and formatting. Of course, thanks to those of you who have been asking these questions over the years, because you truly have inspired more clarification.

A thank-you to Elizabeth Webber for her interest in this work and her talent in illustrating these ideas. My special appreciation goes to Marti and Del Walker, who inspired me to become a teacher. I wish to thank all of our practitioners and students who have spent hours helping with photographs and kept the faith that this book would indeed be produced. Many of these students had the courage to pose in their before-sessions posture for the benefit of your learning. Also, I want to extend my gratitude and thanks to the class photographers: Allison Funk, Bridgett Meyers, Jeanne Pangburn-Roberts, and John Heutmaker, as well as Roz Schoenberg, a professional photographer who took certain student photos and my photo.

And, finally, my gratitude to the Stargazer and the vision he holds.

CHAPTER

1

Introduction

Educators and therapists are challenged daily with questions like these: "Why am I getting tense here?" "Why is this symptom occurring?" "How could I have injured my neck while race-walking?" Or they may wonder why a client is not able to feel more balanced, or why one side of the body is overstressed for one person and not for another who is engaged in the same activity.

Numerous factors are to be considered even in beginning to answer some of these questions. The content of this book focuses on body posture: the spatial position of the body segments in relationship to each other and to the whole body. We also will examine the possible consequences of posture on the body's shape. It is not the intent of this book to focus on the body system—cardiovascular, digestive, endocrine, respiratory, lymphatic, musculoskeletal, nervous, reproductive, and urinary. While it is obvious that the health of all the body's systems influences posture, I would like for you to consider how posture may influence all the body's systems.

I have found that the more we who are in the field possess the skills that enable us to collect detailed and accurate information about a client and apply it to help him or her achieve and maintain optimal function, the better we'll be equipped to assist the client to become his or her own teacher and to know how and when to take charge of his or her progress. Generally, I have found that those necessary skills entail the ability to

- observe and assess segmental relationships of the body, and to determine the effects of these relationships on function,
- palpate and discern patterns of hypertonicity and hypotonicity, and to determine whether an increase or reduction of tone is needed,
- develop from the information gathered a hypothesis and session plan specific to each individual,
- use strong communication skills, and select and use language appropriate to each client,
- demonstrate the postural concepts in each individual's body use,
- use refined teaching skills for coaching basic motions, and
- apply the basic motions to complex activities (activities of daily living [ADLs], sports, ergonomics, etc.)

In this book we will focus on the first skill, visual observation. However, there is another factor that many people in the therapeutic and coaching professions are beginning to see as a necessary component in their assessment. That is the ability to observe whole body patterns. Using specific testing to assess an individual body segment tells us everything about that segment in isolation. It reveals little, however, about the relationship of the particular segment to the whole body and the possibility that the dysfunction of that segment is caused by its relationship to other body segments (e.g., shear, compression, rotation, stretch). For example, an ankle does not pronate independently of the knee or the pelvis above it. Every exaggerated spinal curve has a significant effect on leg position and tension. These interrelationships are the reason that we will gain more insight from studying the whole body instead of just the isolated areas of focus.

In practice, we commonly assess patients with multiple complaints, complex histories, and various structural changes in their bodies. We are interacting with the results of many years of movement patterns and trauma, both physical and emotional. If we look at the relationships in the body, then an area of complaint may be only a starting place, a point of departure for assessment. By taking into consideration the functional relationships of the body, we often can trace the dysfunction from the site of complaint backward to important underlying factors. I have developed this understanding of the functional relationships within the body from more than 30 years of teaching thousands of therapists, teachers, coaches, and athletes to analyze these relationships in order to improve the results of their work. It is a perspective that represents a new paradigm for observing human structure and function.

Influences on Posture

There are many influences on our structure and function. Interestingly, the chronological sequence of past events has, in conjunction with all the other influences, determined our specific shape (structure) and the specific ways we move (function). Many times, children fall and hit their heads, badly bruise themselves, or break bones. A child is considered well when the bone mends or the bruise or the headache is gone. However, that single event of injury may have set a postural pattern into motion, leaving one part of the body more off-center to compensate for the area of injury. The child then grows up around the injury instead of growing through that area, inclusively. Subsequent injuries may then lay other compensations on top of the first, and so on.

Usually there is a reason that body parts align themselves as they do. The body shows an innate wisdom as it adopts alignment deviations or compensations in order to maintain its balance. Generally, bodies are very smart and do not make mistakes. They do the best they can to negotiate with their history, injuries, compensations, tasks, and abilities to accomplish an action. However, often the body holds on to these adaptive compensations long after they are needed, resulting in an overall loss of full potential and freedom of movement.

I have found that the body is quite resilient in its ability to correct itself from the adverse effects of compensating, given the appropriate formulas for neutralizing the pattern. It is rewarding when, as therapists, teachers, and coaches, we are able to facilitate change, neutralize a negative consequence, and help a person move and perform more optimally. It is important to remember that changing one body segment affects the whole system. By stepping back and looking at one part in relation to the whole, our view becomes more comprehensive and clear. When we see the body as an integrated system rather than as many isolated segments, we recognize that we are a complete system in stationary balance and in motion. Being able to see and understand this system greatly enhances our ability to facilitate change.

Consider some of the influences that shape us:

1. genetics
2. in-utero and birth experience
3. developmental phases
4. parental/caregiver patterns of holding, carrying, etc.
5. product designs (furniture, tools, etc.)
6. appliances (braces, glasses, casts, etc.)
7. role models (teachers, movie stars, athletes)
8. fashions
9. activities (baseball, swimming, golf, martial arts, etc.)
10. current age, weight, state of health
11. attitudes, beliefs, personality
12. vocation
13. nutrition, diet, addictions
14. environment (community, altitude, pollution, noise, etc.)
15. physics

In order for therapists, trainers, and coaches to assist someone in changing a pattern, it is necessary to be able to see where that person is in the first place to observe and assess that person's pattern as a point of departure.

Examples

As a physical therapist: If I can see that the center of your trunk is placed posteriorly—a compensation that your body made during recovery from two broken legs—and that it has left your balance precarious because your trunk is placed posteriorly to your foot at heel strike, then I can help your gait pattern by making a correction in your trunk as well as in your feet.

As a massage therapist: If I can quickly assess your overall body pattern in alignment, then I can correlate that information with what I feel in your musculoskeletal tissues and better design a session for overall improvement.

As a personal trainer: If I can see how your current alignment pattern contributes to your neck and shoulder tension when you lift a weight, then I can begin to help you change that pattern throughout your whole body in order to avoid being overstressed.

As a psychotherapist: If I can see that your current alignment pattern no longer matches your current expression, but is a flexed-ribcage posture that once matched feelings of grief, then I can form a strategy to help you.

As an occupational therapist: If I can see your body postural pattern, I can teach whole-body usage for improving your ADLs.

As an ergonomics consultant: If I can see your pattern of lumbar hyperextension in relation to hyperflexion in your thoracic curve, in relation to your being seated in a low chair, I can suggest changing your whole-body use to assist the isolated area of complaint (thoracic outlet).

Goals of This Book

People often complain about an isolated area: low back, shoulder, neck, knee, headaches, and so forth. Many times, therapists are asked to work only on an isolated area of the body. Many therapists trained in Aston *seeing* skills find it rewarding to be able to see quickly and accurately the relationships between the involved area and the contributing areas, and then be able to accelerate overall improvement while reducing the specific symptom.

Many people whom I have trained have consistently commented on three things about Aston *seeing* skills that are of particular value to them. First of all, the Aston Paradigm provides a different perspective on neutral alignment. Second, Aston *seeing* skills help in understanding how the dimensions of each body part interplay with segmental alignment. Third, and most significant, *seeing* skills help therapists attain an understanding of the relationships among the parts of the body that can give tremendous insight into how the adverse effects of physical and emotional trauma can be compensated into a dysfunctional pattern. Whether you are a novice or experienced therapist, teacher, coach, or trainer, or a student in these arts, it is my hope that you will find this approach to be fresh, inspiring, and of great practical use in your work.

This workbook has been created to provide a comprehensive self-teaching tool for developing *seeing* skills. These include 1) accurately assessing the parts of the body in three-dimensional space (including consideration of alignment and dimension, or volume of segments) and 2) understanding the relationships within the body that cause the specific alignment and dimensions we observe. Contained in this book are tools to enable you to observe a client's body and quickly identify whole-body relationships that may contribute to an area of interest or complaint. Chapter 3 will utilize these skills from a different perspective of the Aston Paradigm.

This workbook offers: 1) experiences to develop and enhance observational skills, 2) notation techniques to assist easy visual recall for recording observations, and 3) a process for thinking about the significance of observations. These elements are the precursors to therapeutic choices such as the work sequence, modalities, and areas of focus. Detailed information on designing treatment plans based on your visual assessments is outside the scope of this book; however, I anticipate that you will find many applications for the information included here.

The Aston Paradigm has spawned a new perspective for evaluating musculoskeletal function—an evolution in movement education, fitness, product design, massage, and bodywork—and a new set of ergonomic guidelines. It is because of the positive feedback I have received about the value of this information, from health care and fitness professionals who have been trained to use these techniques, that I have created this workbook.

Beyond the scope of this book is an entire system of movement education that has grown from these ideas, as well as numerous product designs. I have included a few examples of applications for your consideration and exploration.

How to Use This Workbook

The exercises and examples given in this book are designed to impart information and to exercise and develop skills. This is a hands-on approach that will enable this two-dimensional medium to communicate more accurately the three-dimensional reality we are exploring. I encourage you to take the time to work through all the exercises.

Because the advanced material is a development of the ideas and specific examples given over the course of the book, advanced and beginning students alike should go through the book from beginning to end the first time through. Although you may have experience in analyzing alignment, this new paradigm uses different landmarks—a different view of optimal alignment. Going through step by step will enable you to become familiar with this new perspective.

The material is arranged from simple to complex relationships. Each section contains the repetition of a skill through a variety of examples. This is based on a learning model that suggests that information becomes more fully and easily integrated when one masters a skill before moving on to a new concept or skill. Therefore, certain pieces of information will be presented in isolated, two-dimensional, or oversimplified forms at first. Each piece will be woven into the complex framework of the three-dimensional reality of the body by the conclusion.

Generally speaking, each section includes text, examples, exercises for applying the ideas, and some implications and clinical examples or applications. You may want to go through the book the first time just working with the development of skills. Then go through again and include the application of the concepts to practice. You might make this choice if you have limited clinical experience. Others may find it most useful to see the implications for practice as they build skills. At any rate, you have the option to go through the entire text using just one of these tracks, or combining them in any way you choose. Each way is a different experience, and it may be useful at different times. In this way, the content expands as your experience increases and questions change. I have designed this book so that you may write in it. The skills it teaches require practice so that you can use them with ease.

In the exercises, I have suggested that you make marks on the drawings and photos. For those of you who prefer not to mark in your book, I suggest that you purchase about three transparencies and a china marker or a dry-erase marker to enable you to do the exercises, erase, and move on. You also may want to use a 6-inch length of dental floss or string to act as a plumb line.

Many photos are provided as examples within the text. To expand your opportunity to practice, I recommend that you study photos of clients and friends, figures in textbooks, pictures in magazines, and photos of your own body. When possible, it is good to have four views: front, right side, back, left side. The client should be wearing clothing that enables you to see body contours.

We will use two general parameters for observation:

1. Alignment—the placement and the relationships of body parts to each other and the whole body with respect to a plumb line.
2. Dimension—the internal volume or mass of body segments or of the body as a whole. The amount and placement of the three dimensions of length, depth, and width have great significance for function.

Following the simple-to-complex format, we will begin assessing these two categories separately, and then see how they affect each other.

Chapter 2 presents simple steps for observational assessment and Chapter 3 introduces Aston-Mechanics℠. Chapter 4 contains tips on how to integrate these principles into your practice, and Chapter 5 offers exercises on applying what you have learned.

A glossary at the end of this workbook contains terms with which you may be unfamiliar.

Pretest A

In order to appreciate your progress while working through this manual, you may find it helpful to record what, and how, you are *seeing* now for comparison later. Take a moment to look at the following photos and write down what you see. Use an extra sheet, if necessary.

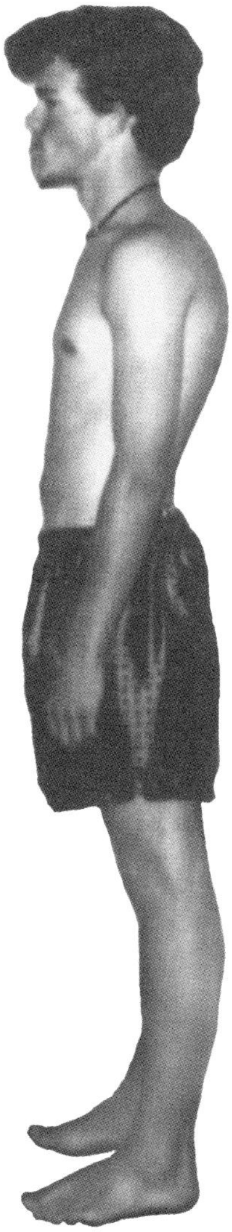

Answer these questions.

1. What do you see?

2. What do you think this person's complaints might be?

3. What areas might you address in a session and why?

Notes

Pretest B

For your learning and personal application, have a friend take a set of Polaroid™ photos of you from different views:

1. front

2. left side

3. back

4. right side

5. as you might stand while waiting in line

Tape your photo series here and make the same observations as in Pretest A.

CHAPTER

2

Alignment

Introduction

Alignment is the placement of the body segments in space in three-dimensional relationship to one another as well as to the whole body. The three planes used for observing alignment are right to left (the frontal plane), anterior to posterior (the sagittal plane) and up to down (the transverse plane).

Being able to assess the alignment of the whole body helps us understand primary and secondary contributions to a particular individual's symptoms. Sometimes it is easy to imagine that all people who have the same symptoms share the same body patterns of alignment and might therefore benefit from the same treatment. These three people have low back pain, which is a common complaint.

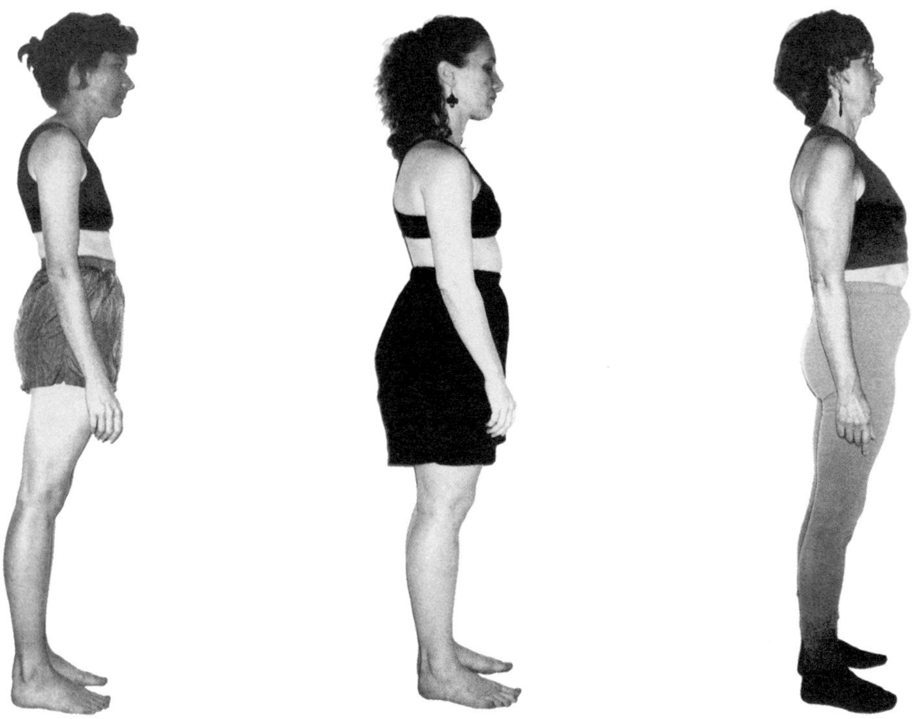

Observe the very different body patterns shown on the previous page. It is important to see that although these people share the same complaint, each pattern adds stress to the back in different ways. A different treatment protocol might be most helpful in these different cases.

Note: The majority of the photos in this book show the clients and students demonstrating their "before" session photos. While this might exaggerate their patterns, it also makes it easier for you to learn.

Observe the following common patterns of alignment.

1. Suck-and-Tuck position

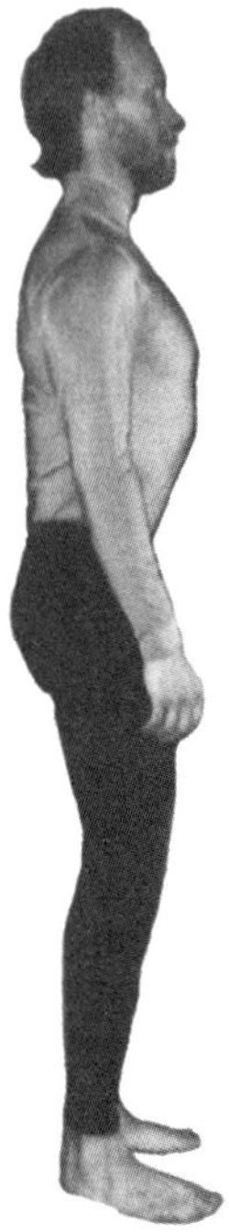

2. Dancer position (extension)

3. Slouched posture (standing)

4. A dentist's position (rotation)

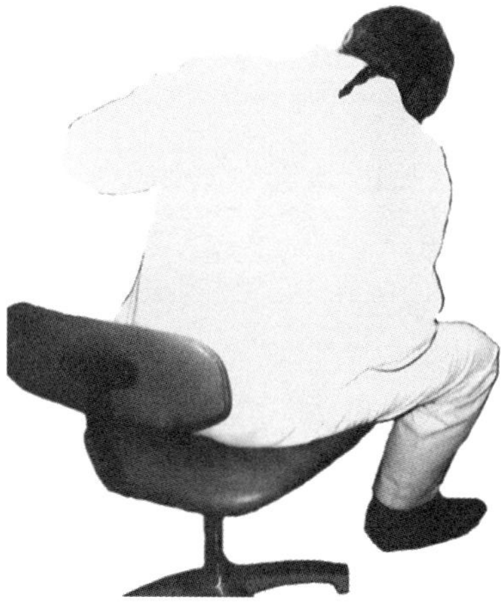

Many of these patterns have been learned in order to achieve a specific result. Although they may be effective for the desired goal, there also may be certain unwanted consequences elsewhere in the body. Some of the adverse consequences are demonstrated below.

1. Suck-and-Tuck position

Action	Purpose	Consequences
Suck in abdomen Tuck pelvis under	To stabilize the spine To improve posture	Possible compression and a decrease in the range of movement of the hip and spine

2. Dancer position (extension)

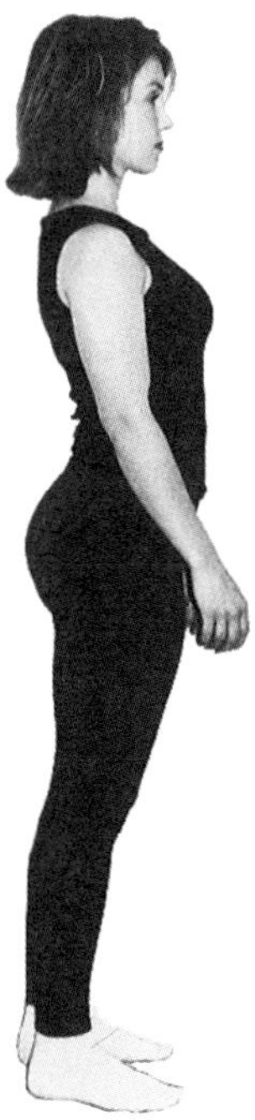

Action	Purpose	Consequences
Lift up chest	To elongate the aesthetic line To lift off the legs	Stretching and a decrease in shock absorption through the spine

3. Slouched posture (standing)

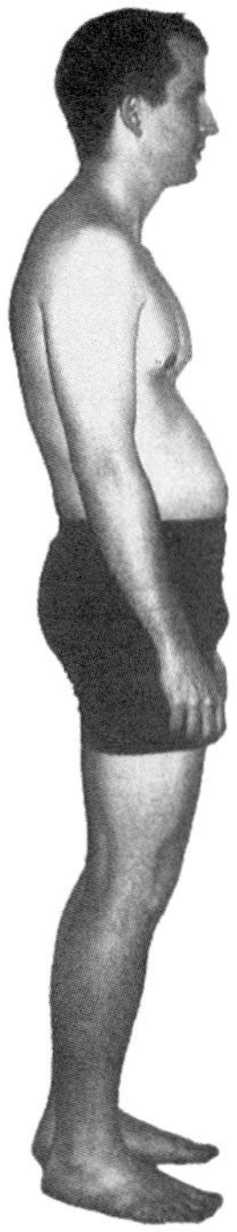

Action	Purpose	Consequences
Relax the trunk	To let go, be casual, relieve stress	Compression in the chest, decreased breath, increased pressure in the abdomen

4. A dentist's position (rotation)

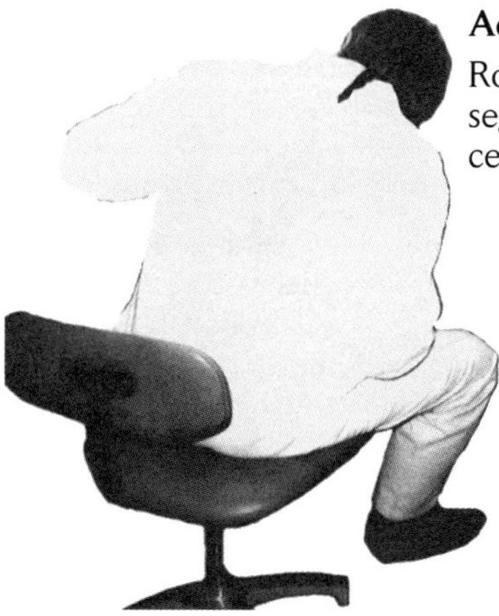

Action	Purpose	Consequences
Rotate body segments around center of body	To negotiate the patient, chair, and task	Bending without adequate support, prolonged, uneven stress on right and left sides of body

Clearly, each isolated action influences the whole body pattern. This workbook will provide a framework for understanding the body as a whole system.

Planes

We will assess alignment in three planes that are defined here in technical terms. In the rest of the text, we use more ordinary terminology.

Sagittal Plane (side view)—A vertical plane extending from front to back, dividing the body into right and left halves. This plane is viewed from the right or left side of the body, and describes the anterior/posterior relationships. We will refer to alignment in this plane as being anterior or posterior, forward or backward, to the point of reference.

Frontal Plane (anterior/posterior view)—A vertical plane dividing the body into front and back, viewed from the front or back. Medial-to-lateral relationships are seen in this plane. We will refer to the alignment in this plane as being centered to the right. We will refer to alignment in this plane as being to the right or left of the point of reference.

Transverse Plane (upper/lower view)—A horizontal plane dividing the body into upper and lower parts. This plane is viewed from the front, back, or side, and describes the up (higher) to down (lower) relationships. We will refer to one part as being higher or lower in relation to the point of reference.

We will look at each of these planes separately and then combine them to have a simultaneously three-dimensional (3D) description of the alignment. This is important because any real analysis of a 3D object must include all three planes. Consequently, if an optimal change to a body segment occurs, we need to take into account its effect within all three planes because the body is three-dimensional. We must consider that a simple change in one plane in one body segment also will affect all other segments of the body in all three planes.

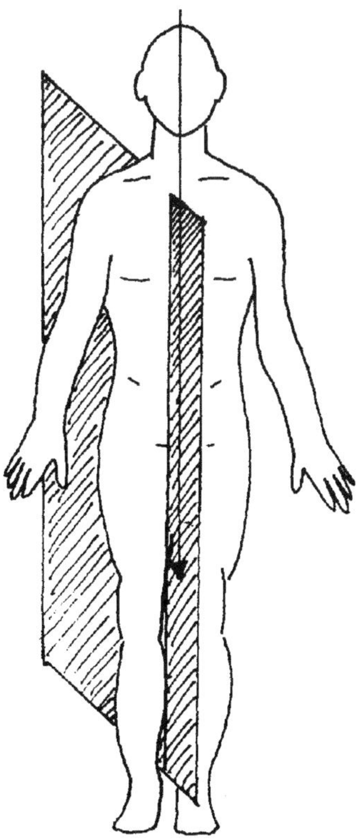

Sagittal

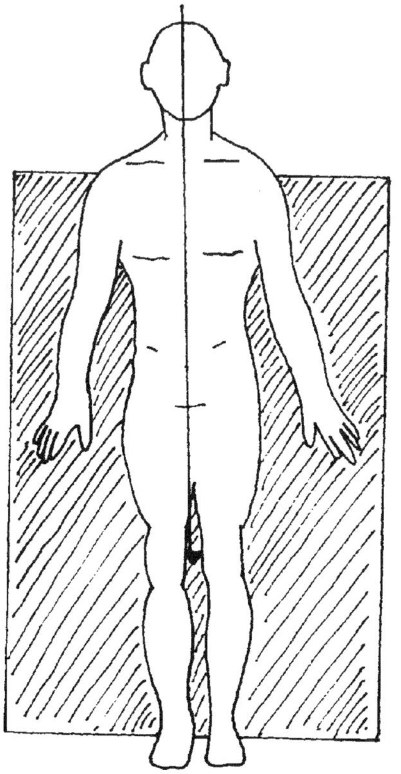

Frontal

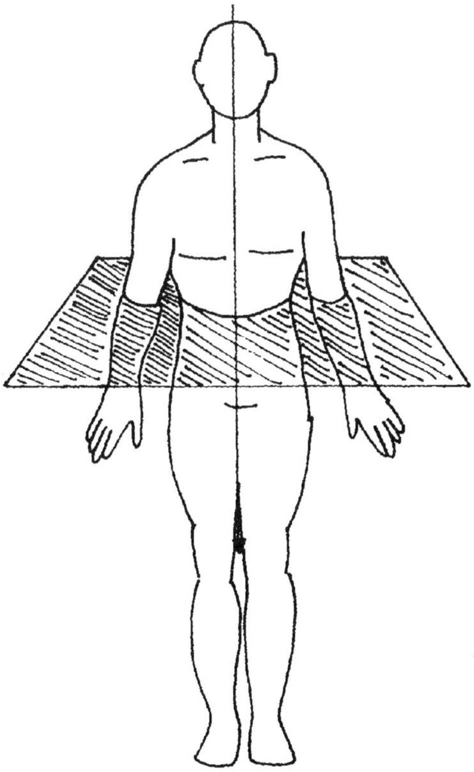

Transverse

Shapes

Take a quick look at these shapes. Notice that they have different designs and sizes.

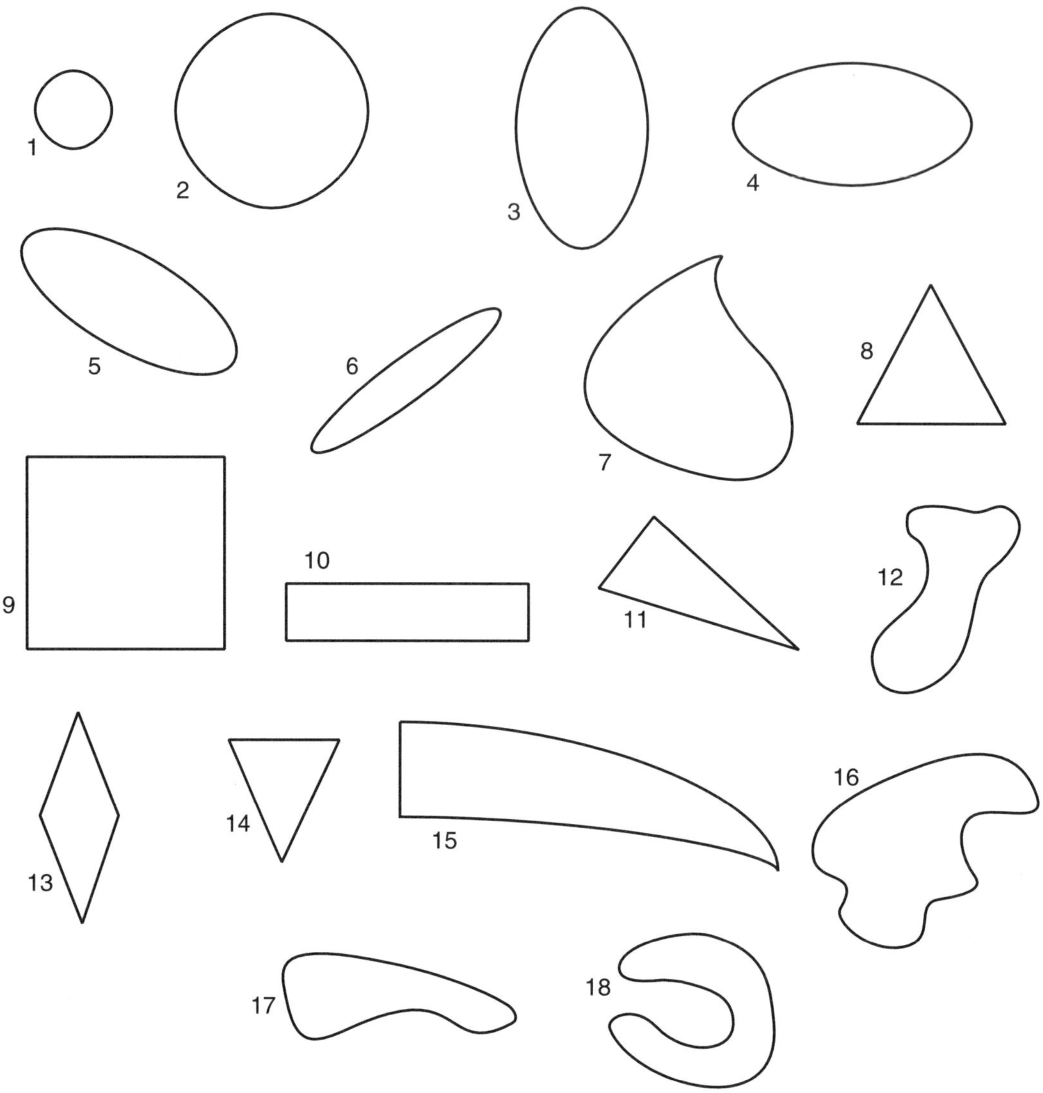

Identifying the Center

When an object is symmetrical, its center of gravity (COG) is the geometric center of the object. When an object is asymmetrical, its COG is toward the heavier end, at the point around which the mass is evenly distributed. Look at each of these individual shapes and determine the location of the center of each segment.

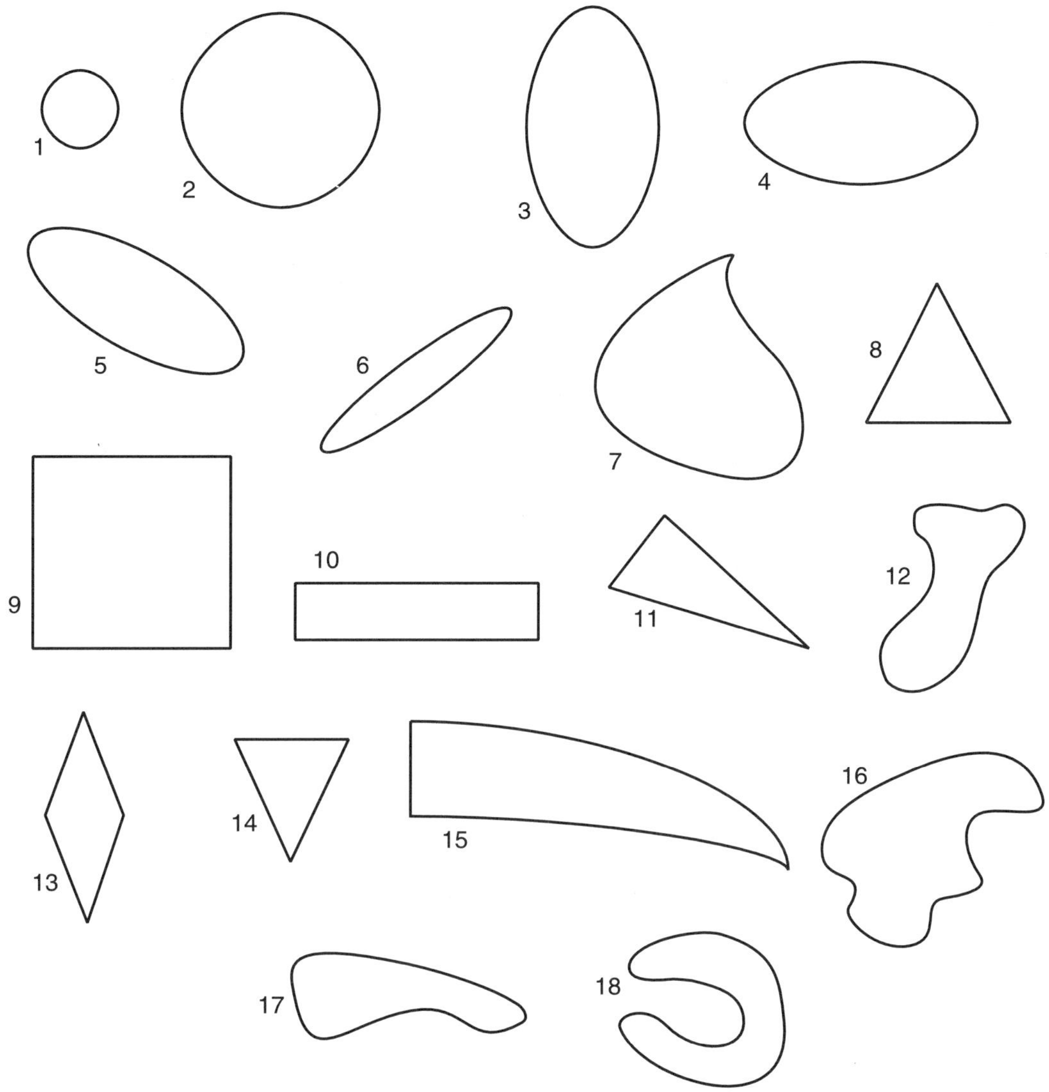

Exercise 2.1

As you glance at the shapes, you will begin to get a sense of how visually to find this center point. Place a transparency over these shapes and, with a wax pencil, locate and mark the center of each individual shape.

Placement

In examining the alignment of body segments, we first look at the placement of the geometric centers, then at the relationship of the segments or geometric centers to each other. We could say that the placement of one part is right or left, up or down, anterior or posterior, to another.

Exercise 2.2

In these irregularly shaped objects, consider the placement of the geometric centers in relation to each other.

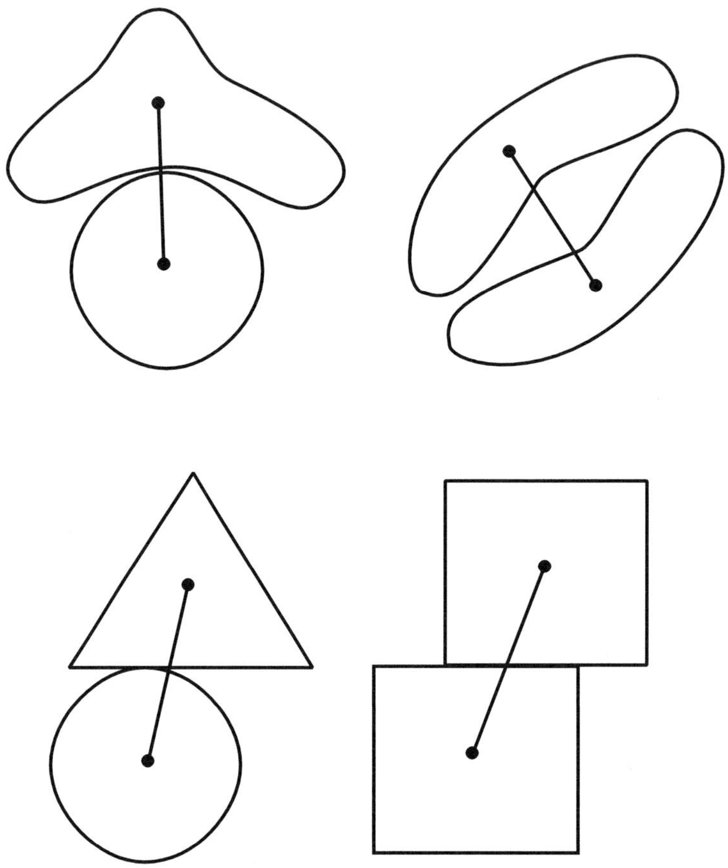

Connect the center points of these pairs of shapes.

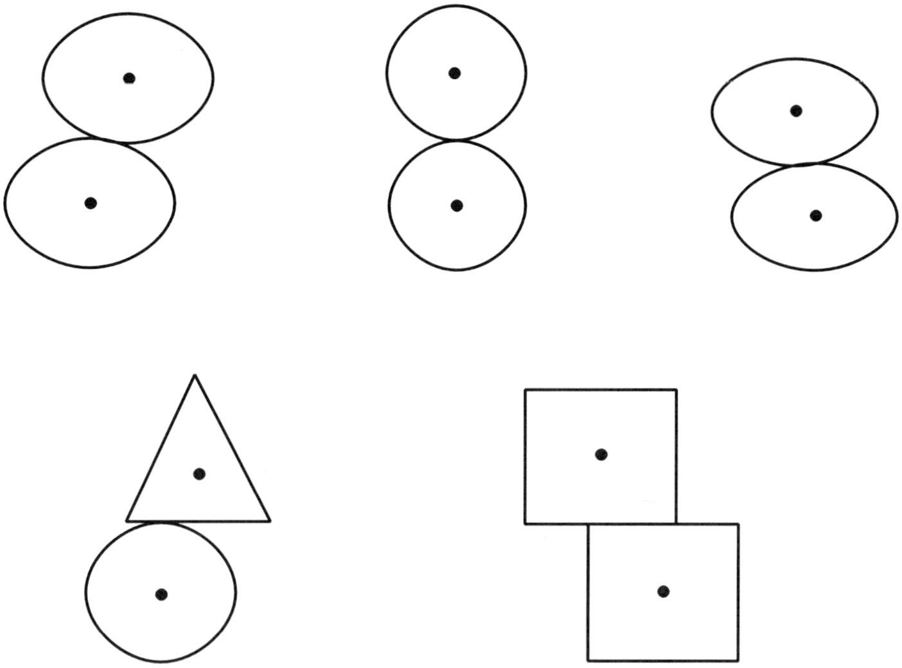

And these…

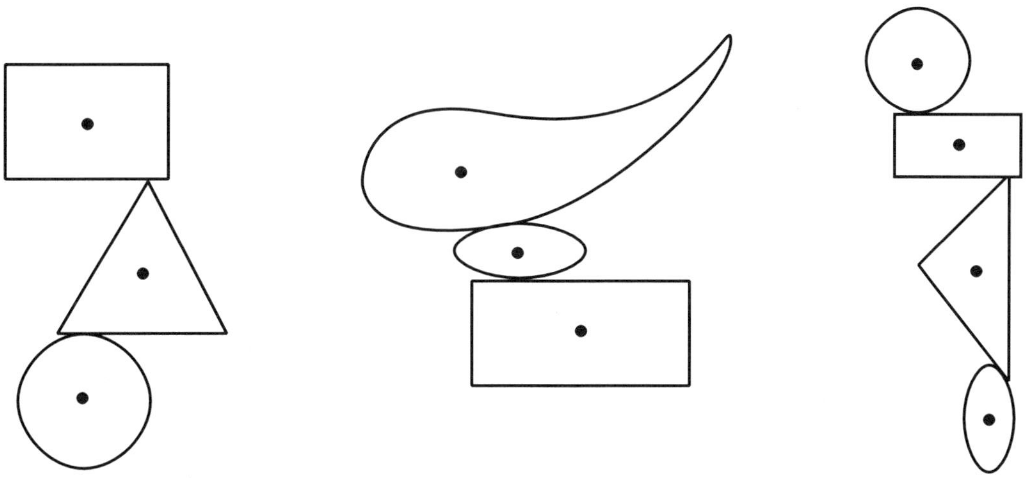

Draw the lines on a transparency and then look at the lines by themselves.

Forces

To understand the effect of body segments on one another, you first must understand the influences of forces. This may help you understand the importance of being able to see alignment patterns in your clients. Once you are able to identify the alignment pattern, you will understand which force is affecting the skeleton and soft tissues. This may give you a better understanding of your clients' pattern and how improving the alignment, and therefore neutralizing the forces acting on their structure, will decrease their symptoms.

When forces act on the body, they create forces within the body's tissues. These are called stresses. Often, the affected segment changes in length, depth, width, or shape (deformation) because of these stresses. This is called strain.

Gravity

Gravity is a continuous force that affects each body. We know that gravity acts as an attractive force between masses. The mass of the human body being smaller than the Earth's mass means we feel a continuous pull toward the center of the Earth. It is the force of gravity that gives the body weight.

Example

Shape 1 is being pulled down onto Shape 2, and the consequence is, in effect, to give weight to Shape 2.

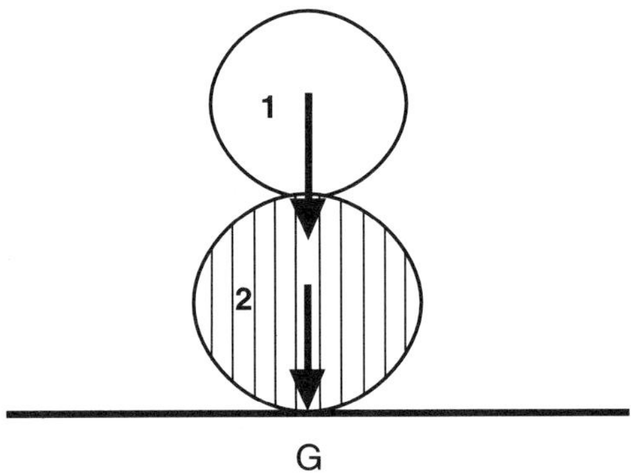

Ground Reaction Force (GRF)

There is an opposing force to gravity known as ground reaction force (GRF). It is the opposing force we experience when we push against the ground (or any other surface). Newton's third law, the law of reaction, tells us that forces come in pairs. Gravity's pair is GRF. This force is in direct proportion to the effort we apply as we push: For every action, there is an equal and opposite reaction.

Our interaction with these opposing forces can determine whether our body segments work effortlessly with one another or whether we use increased effort. We also can use GRF to our advantage to help decrease compressive forces in the body and maximize the body's internal volume. When gravity interacts with GRF, as in Newton's law, the two forces are equal in magnitude and no motion occurs. This implies that the body is static. In the Aston model, we add the element of varying time to this equation. Because the body is continuously moving, one can consider these dynamically alternating forces, and students can be taught how to recycle one force into another. The interplay of gravity and GRF increases the opportunity for the positive use of intermittent compression and distraction in which the joints are lubricated and the nutrients are exchanged.

As students and patients are introduced to the GRF, they are able to neutralize many of the negative effects of gravity on the body. Another way to utilize GRF to our advantage is by pushing with intention down into the surface on which we are supported. The resultant GRF will increase. If we push with more effort into a hard surface, we show little displacement down because of the rigidity of the surface, but because of the increased effort downward, the opposite GRF upward is increased in magnitude. Because the upward force is not resisted by a hard surface like the ground (we are pushing up into air), the net effect is that there is a movement upward, or a lengthening of the body.

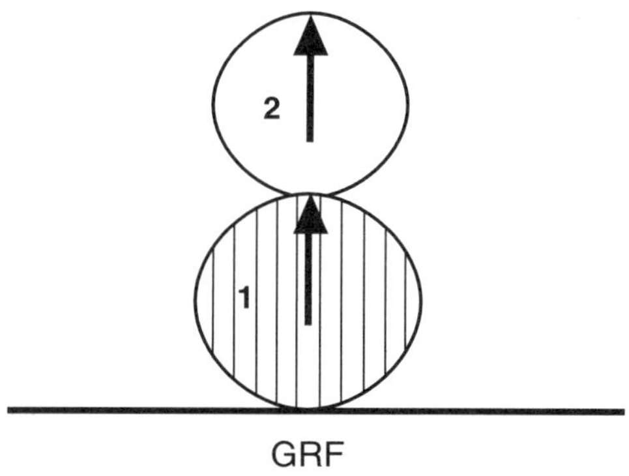

Effects of Gravity and Ground Reaction Force on Three-Dimensional Shapes

Simply looking at these objects in relation to each other, we see that some are more directly aligned and others are at a diagonal relationship. In gravity, the more direct alignment obviously translates to easier balance. Before you apply these ideas more specifically to bodies, think about some of the effects that one object will have on the other, given different alignments.

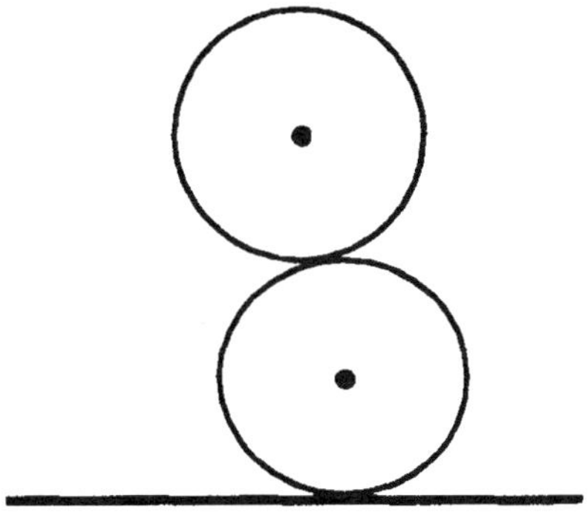

Geometric Centers

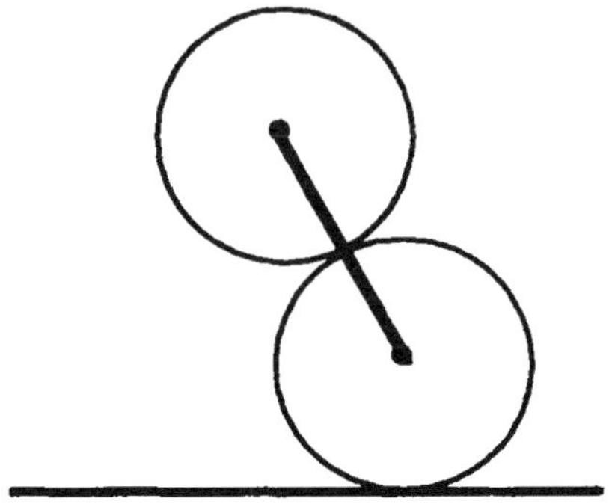

Connecting Centers Creates a Diagonal Line

Let us examine the direction of the effect of gravity and GRF.

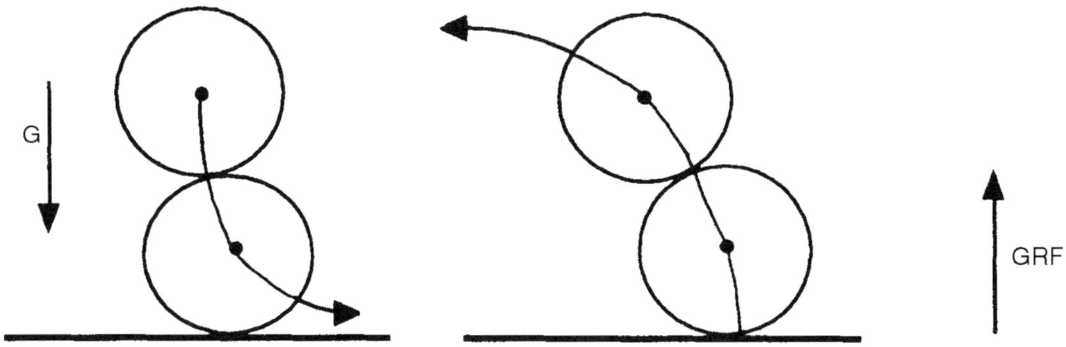

Imagine that these shapes are three-dimensional. Look at the shapes with attention to the direction and influence from gravity and GRF.

The combined effect of gravity and GRF

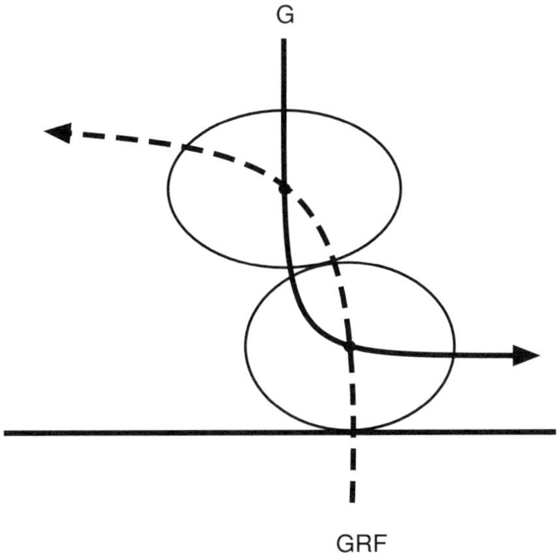

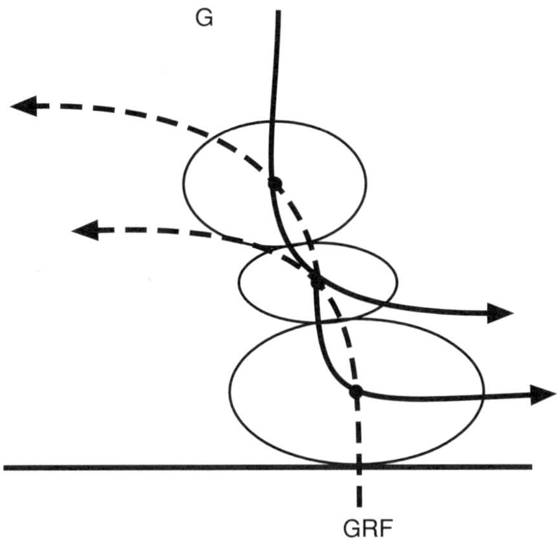

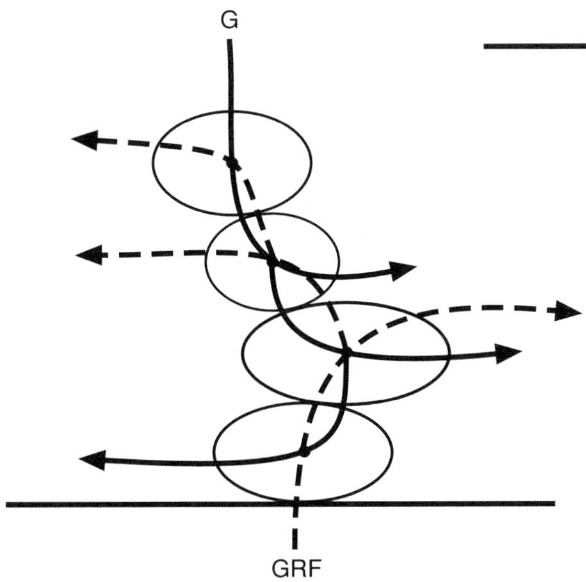

If all the segments of the body were in "perfect alignment," the force of gravity would run directly down through each center of mass and each segment would directly weight the segments below it. This is not often the case. It is important to note that as a segment is shifted out of alignment, the displaced force of gravity is no longer purely vertical but now has an additional direction of force. Now the segment will have additional forces to contend with besides gravity and GRF. These are compression, tensile, shear, and rotational forces. These forces are exerted on our musculoskeletal system with many consequences.

If you can imagine these body segments as spheres, you can see that every segment's misalignment affects the segment below it. If the segments were merely spheres, the balls would roll in the direction of the arrows. But the segments of our body do not roll away. They absorb some of the force of gravity by weighting the shape in the direction of the arrow. This creates forces that we will now discuss.

Compression Forces

If the forces acting on the body are toward each other, acting along the same line, they create compression. The compression stress results in strain (deformation), which shortens and widens the structure.

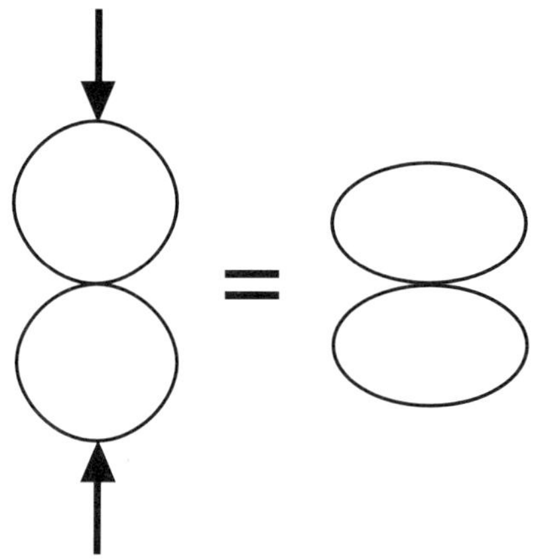

Tensile Forces

If the forces acting on the body are in opposite directions and acting along the same line, they create tensile or distractive forces, or a stretch. The tensile stress results in strain (deformation), which elongates and narrows the structure.

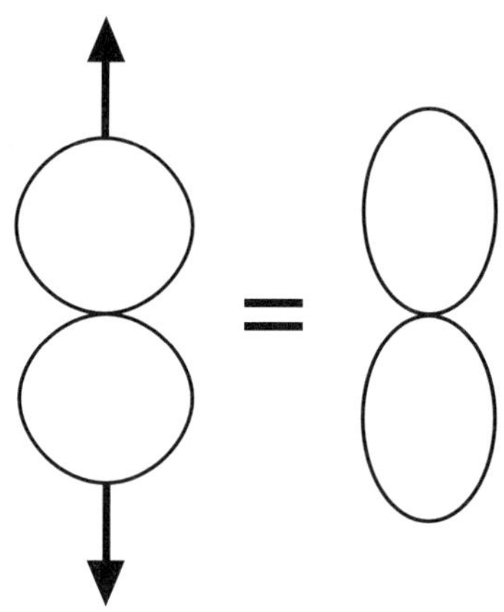

Shear Forces

If the forces acting on the body are toward each other but not in line, they create shear. The shear strain causes a positional change in the segments that is in line with the force.

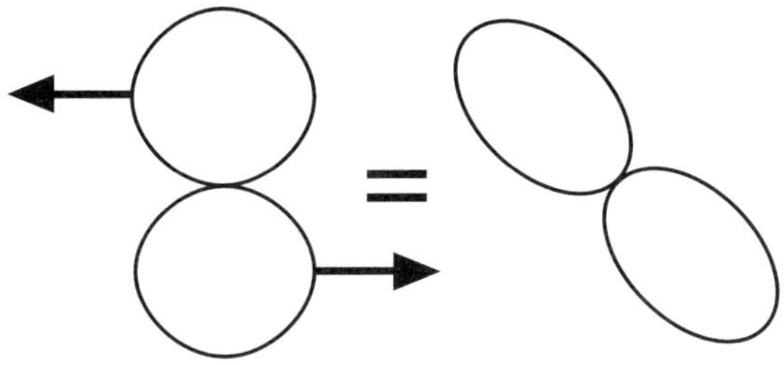

Torsion

If the forces acting on the body twist around an axis, they create a torque or torsion commonly known as rotation. The structure receives a combination of shear, tensile, and compressive forces.

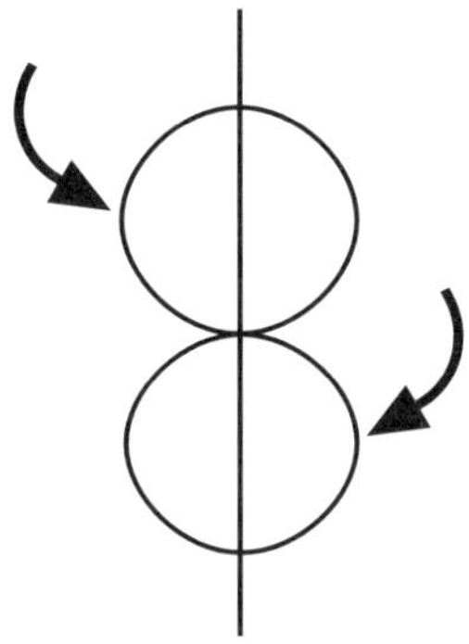

The stiffness, elasticity, plasticity, and strength of our bones, ligaments, tendons, muscles, joint capsules, and connective tissue are tested by the above forces. The change in any segment's shape (deformation) depends on the type of tissue upon which the force is applied and the amount of force. The more resilient the tissue, the less likely it will negatively change shape or be at risk.*

* For more information on biomechanical forces, refer to Nordin, M. & Frankel, V. H. (1989). *Basic biomechanics of the musculoskeletal system* (2nd ed.). Philadelphia: Lea and Febiger and Norkin C. C. & Levangie, P. K. (1992). *Joint structure and function: A comprehensive analysis.* (2nd ed.). Philadelphia: F. A. Davis.

Seeing Alignment in the Sagittal Plane

Body Shapes and Centers

Notice the shapes and positions of the two segments (neck and head) outlined below.

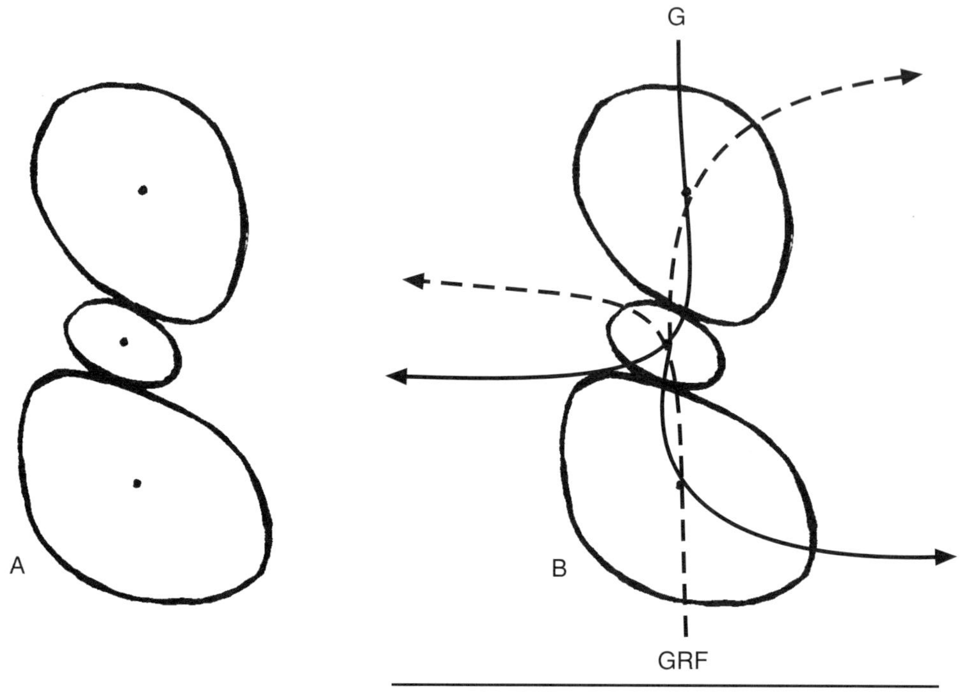

Observe these shapes as an abstraction from the alignment in the photo at bottom right.

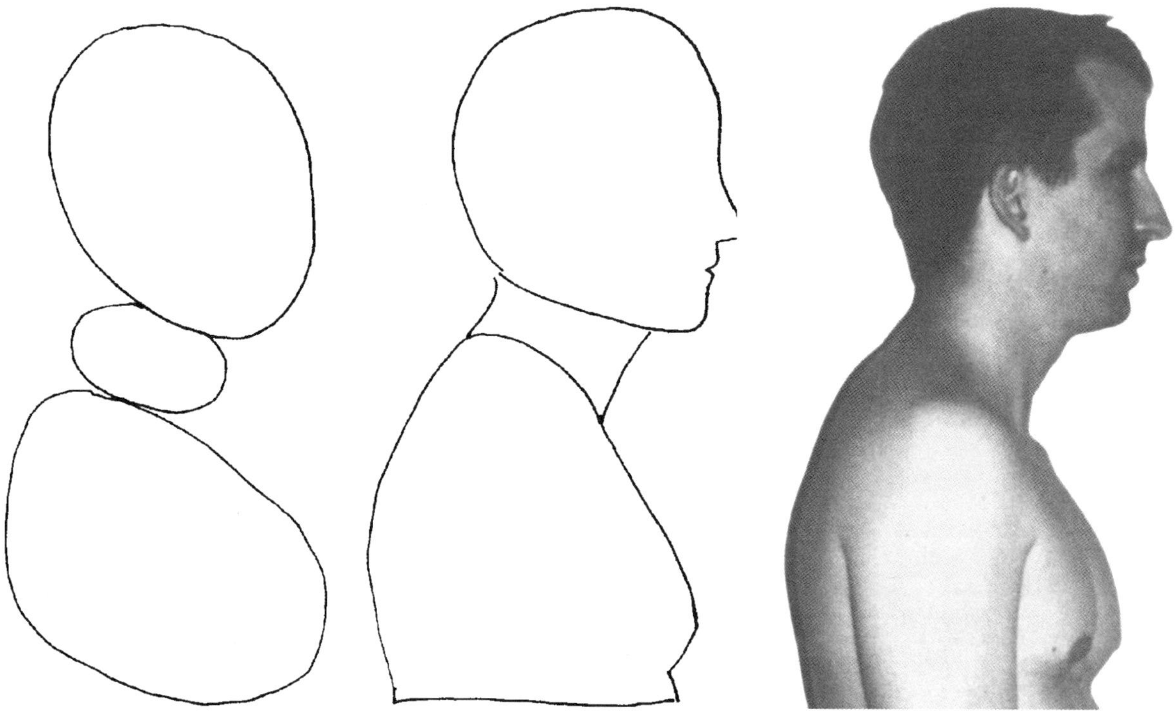

Ball Person and Landmarks—Drawing

Observe the progression of illustrations to see how the following body segments define the ball person:

1. head
2. neck
3. upper chest
4. lower chest
5. abdomen/lumbar area
6. pelvis
7. upper leg
8. lower leg
9. foot

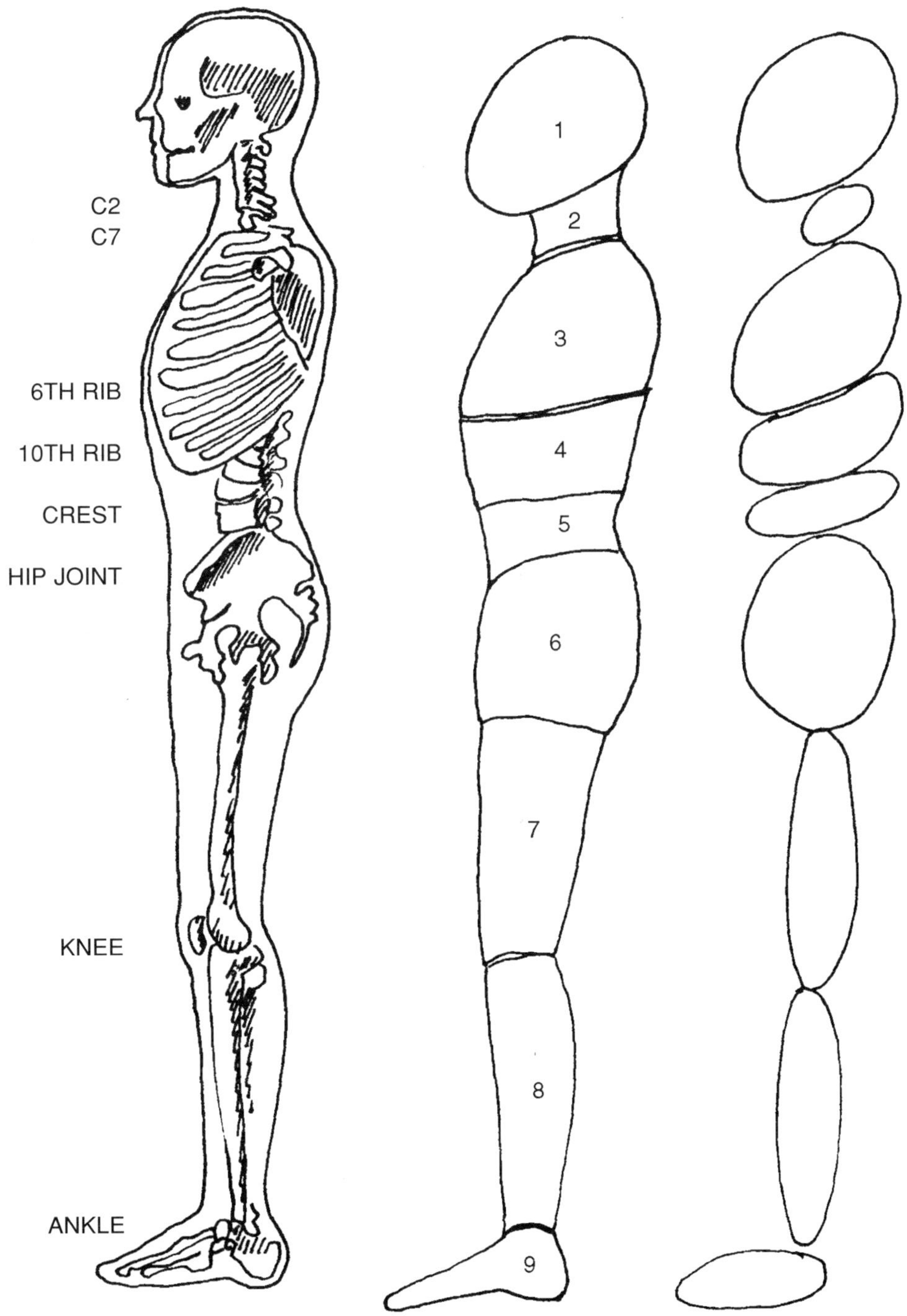

Exercise 2.3

Look at this body shape in the abstract form.

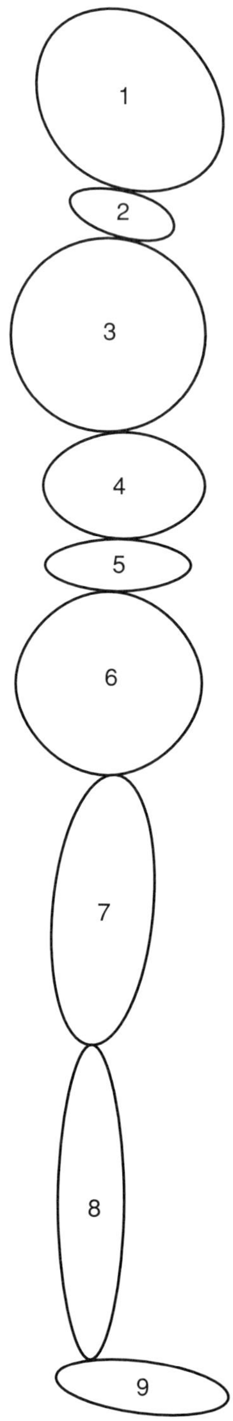

Exercise 2.4

Outline the body segments.

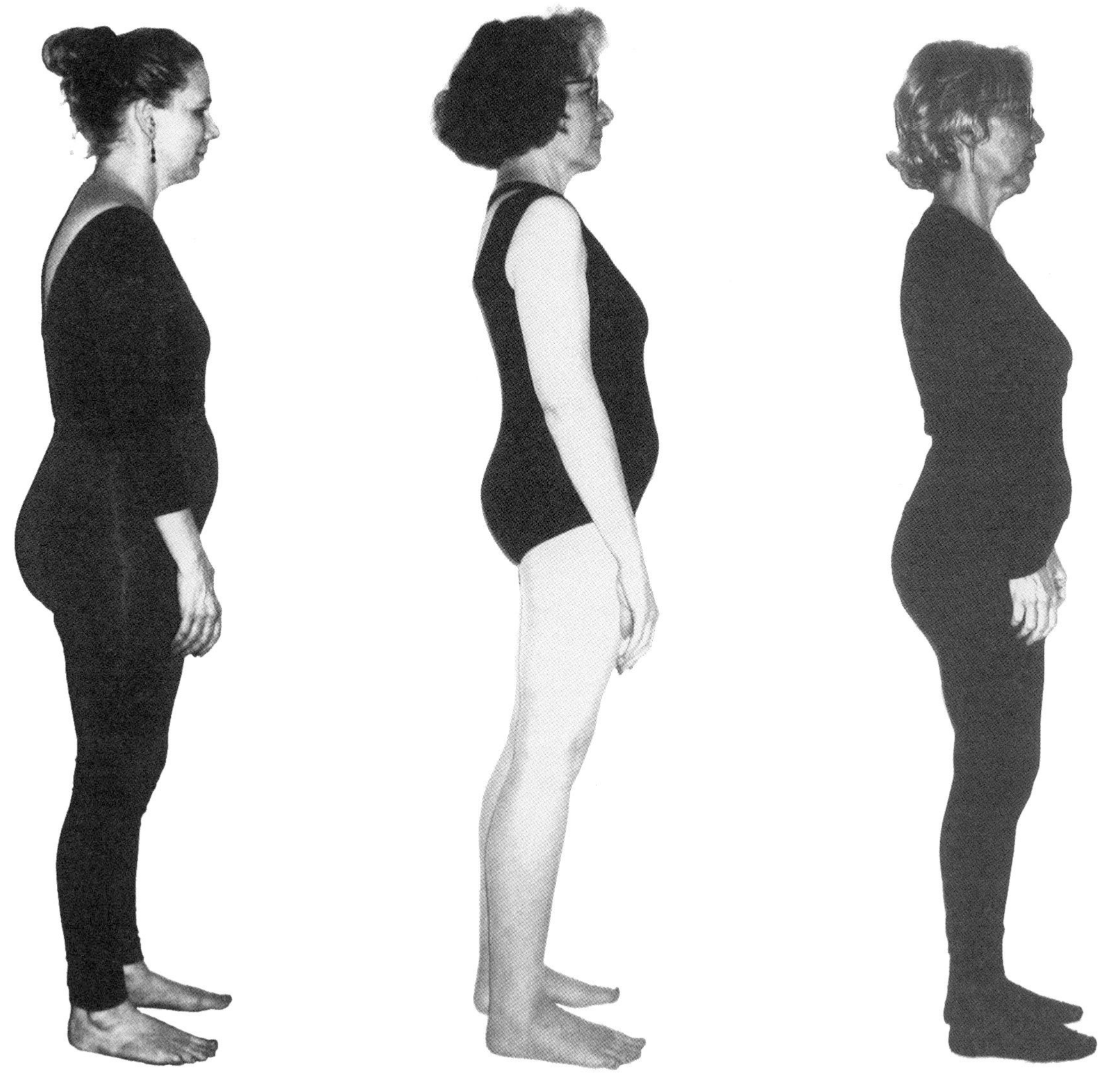

Exercise 2.5

Using the idea of the body segments in abstract shapes, mark the geometric centers of each of the segments below to see the placement of one shape with reference to another. From this point, we will refer to the shift of body segments off the central axis as displacement.

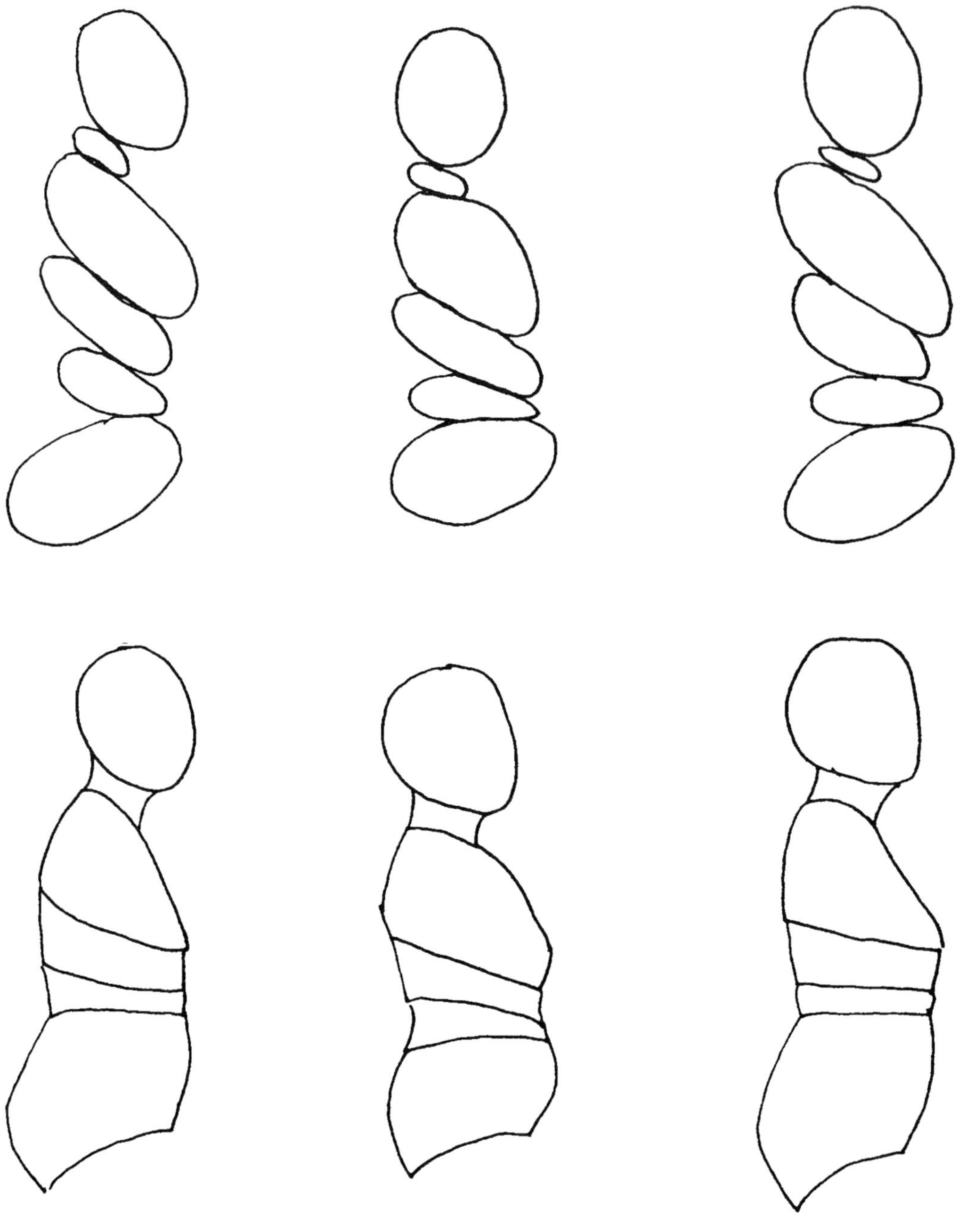

Exercise 2.6

1. Outline the body segments on the following photos.
2. Mark the centers.
3. Connect the centers.

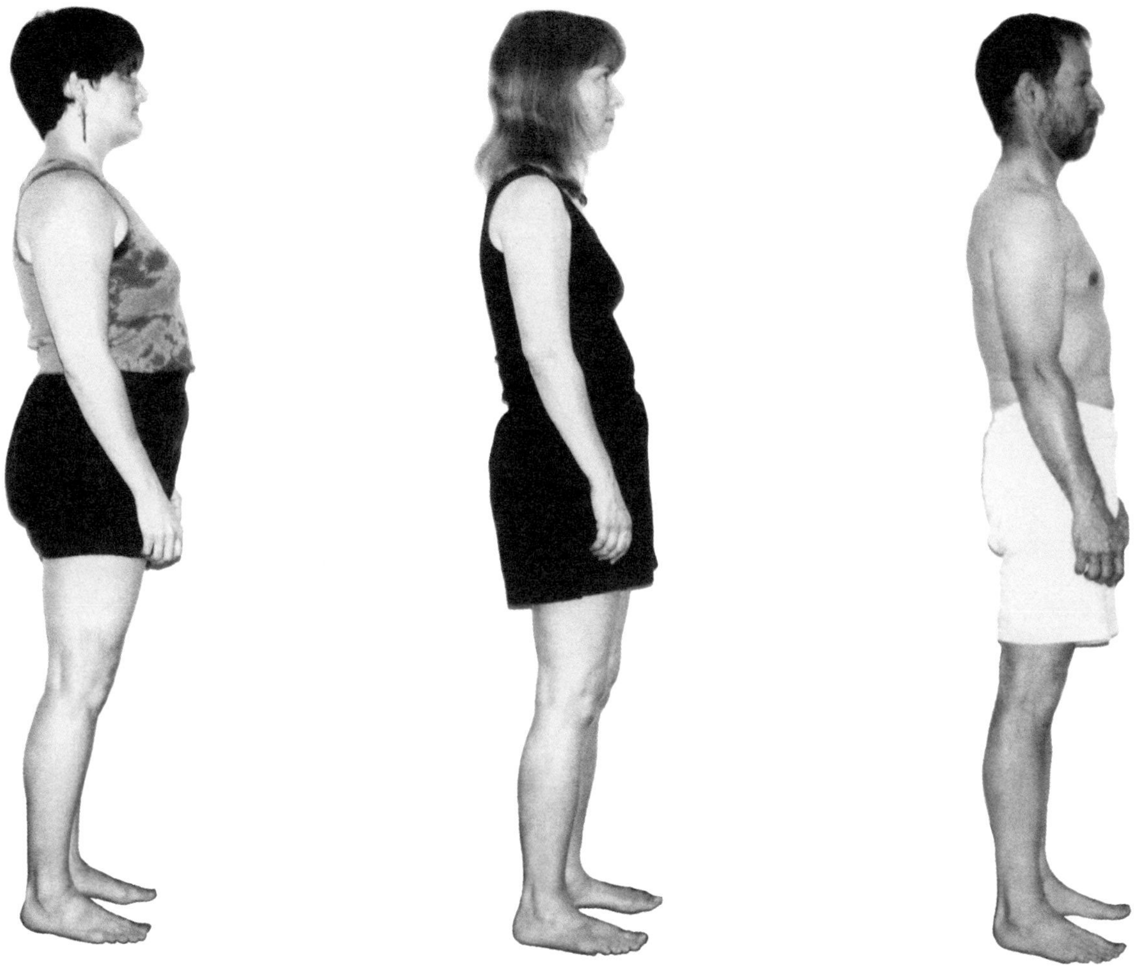

The Effect of Gravity and GRF on Alignment

The previous exercises have given you a point of reference for considering the body as having segments with different shapes, sizes, and weights. The weight of each body segment impacts the others because of gravity and GRF.

For now, put aside the focus on the soft tissue and functions of the body—the muscles, organs, and any ideas about movement, and so forth. This will provide you with the opportunity to gain insight into the way the body segments interrelate with each other.

Exercise 2.7

Consider this abstract body.

1. Mark the centers of each of these segments.
2. Connect the centers to show the relationship of the segments to each other. This indicates the alignment of parts to each other. Make a few mental notes about the directions of gravity and GRF.

Exercise 2.8

Do the same for each of the abstract bodies that follow. Note where the alignment is least direct as well as the consequence to gravity and GRF.

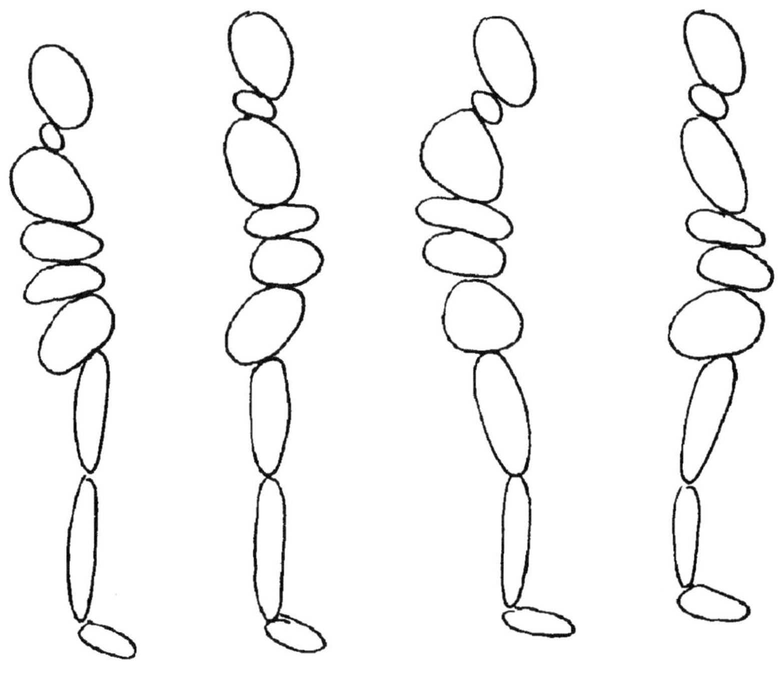

Exercise 2.9

1. Outline the segments on each photo
2. Mark the centers.
3. Connect the centers.
4. Notate the directions of gravity and GRF on the segments.

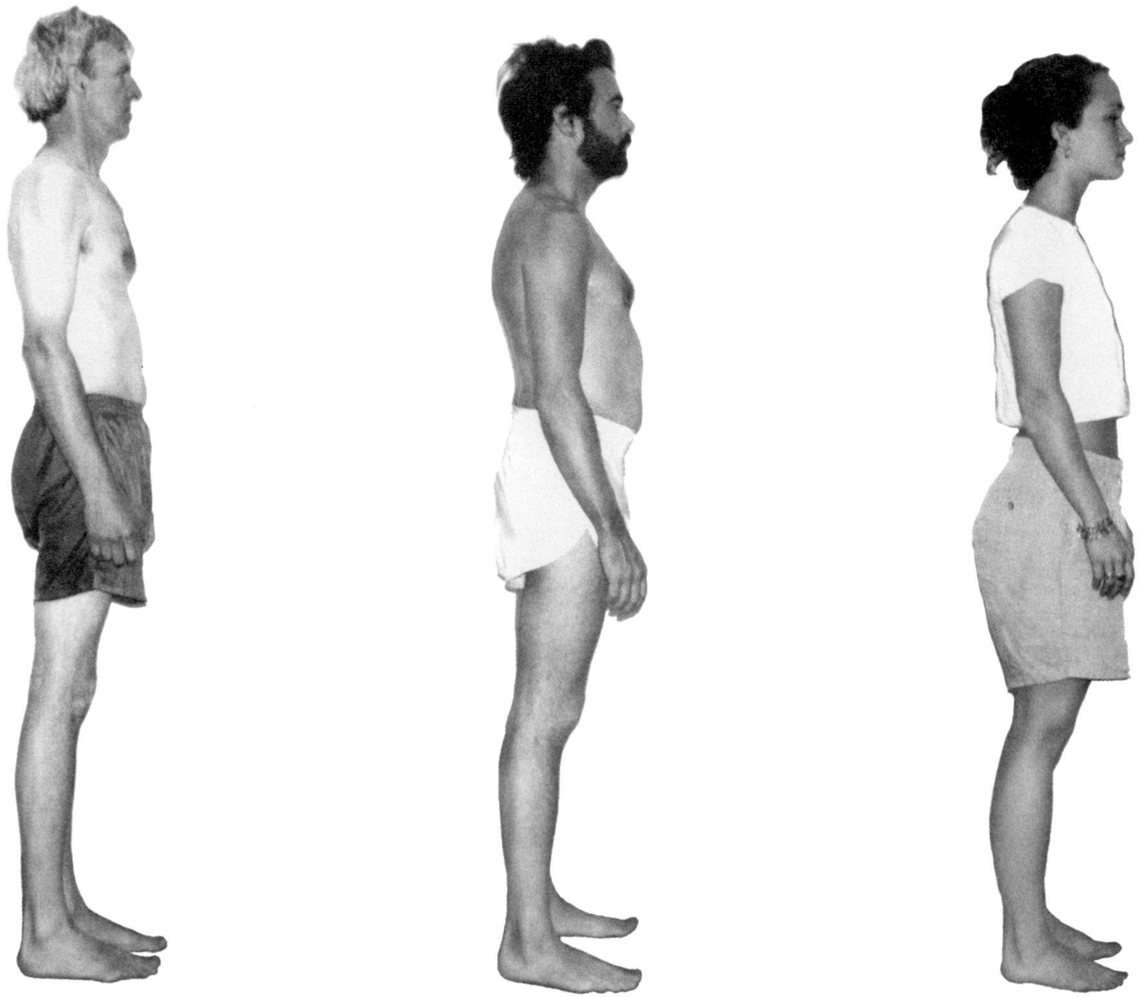

Exercise 2.10

On a transparency, draw all segments of this client's right side. Make your shapes accurate to his true shape, staying within the borders. Refer to p. 29 for the landmarks.

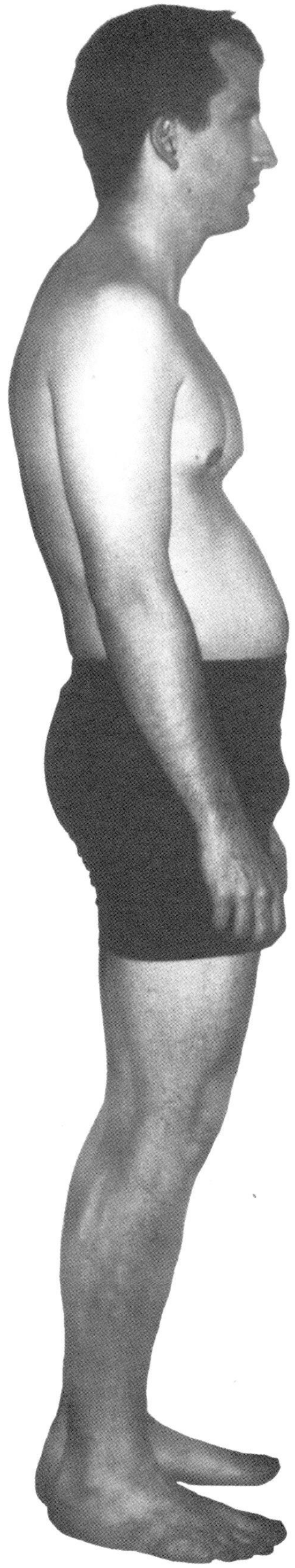

Exercise 2.11

Make an outline of all the body segments, then

1. Mark the centers.
2. Connect the centers.
3. Notate the direction of gravity and GRF.
4. Consider the possible consequence of this alignment.

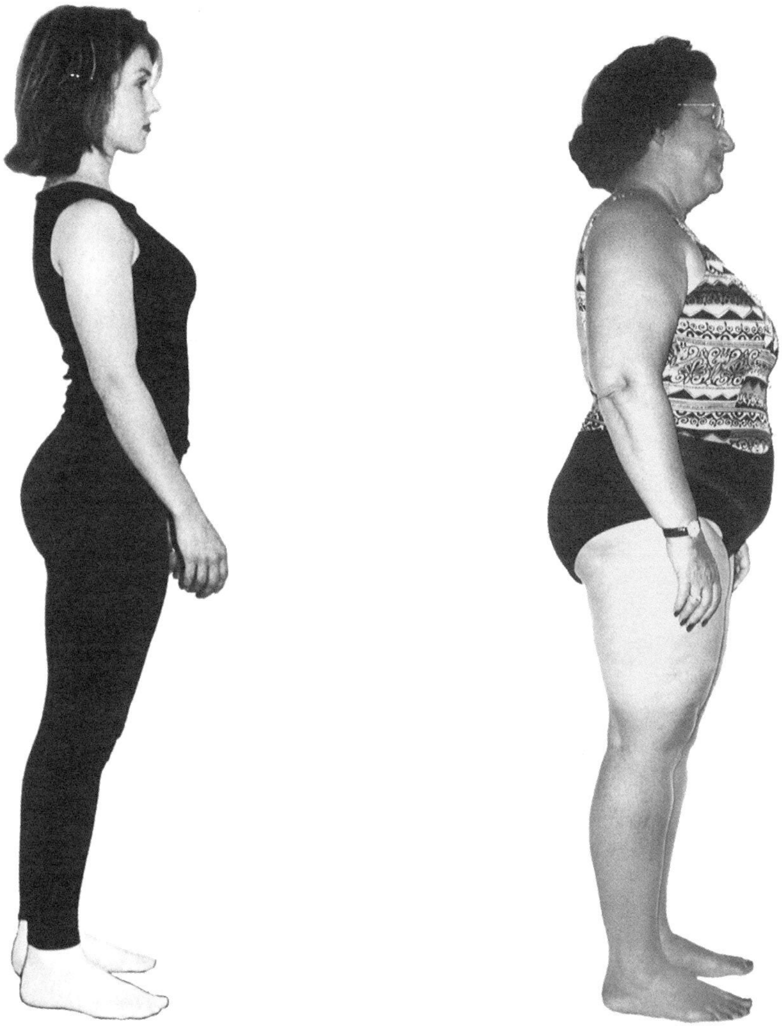

Plumb Line

Center of Gravity

The center of gravity (COG) or center of mass of an object is the point around which all the mass is evenly distributed. It is sometimes thought of as the pivot point or balance point of an object.

The plumb line is used as a tool to assess posture. It is an imaginary line used to create a midline in the body. From this reference line, we are able to observe positional relationships of one body segment to another and to the whole body.

One standard guideline for the plumb line (front view) is a vertical line that begins at the base, midway between the feet and dividing the body into symmetrical right and left sides.

In Chapter 3, we will discuss the differences of the Aston Mechanics plumb line and its use to identify the dimensional aspects of each body segment.

Plumb Line—Side View

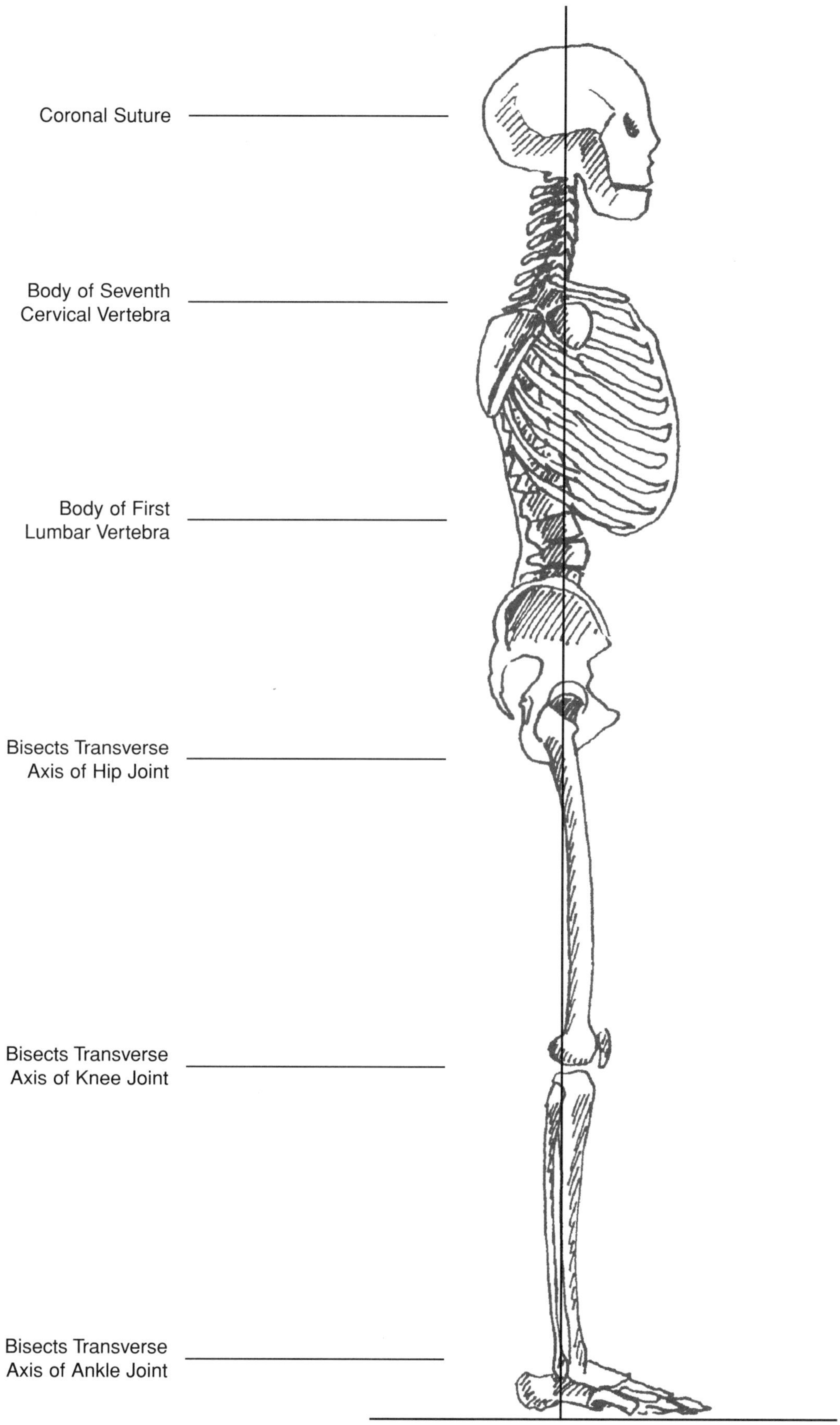

Ideal Plumb Alignment

Exercise 2.12

In the examples below, draw the standard plumb line, mark the centers of the segments, and notice their relationship to the plumb line.

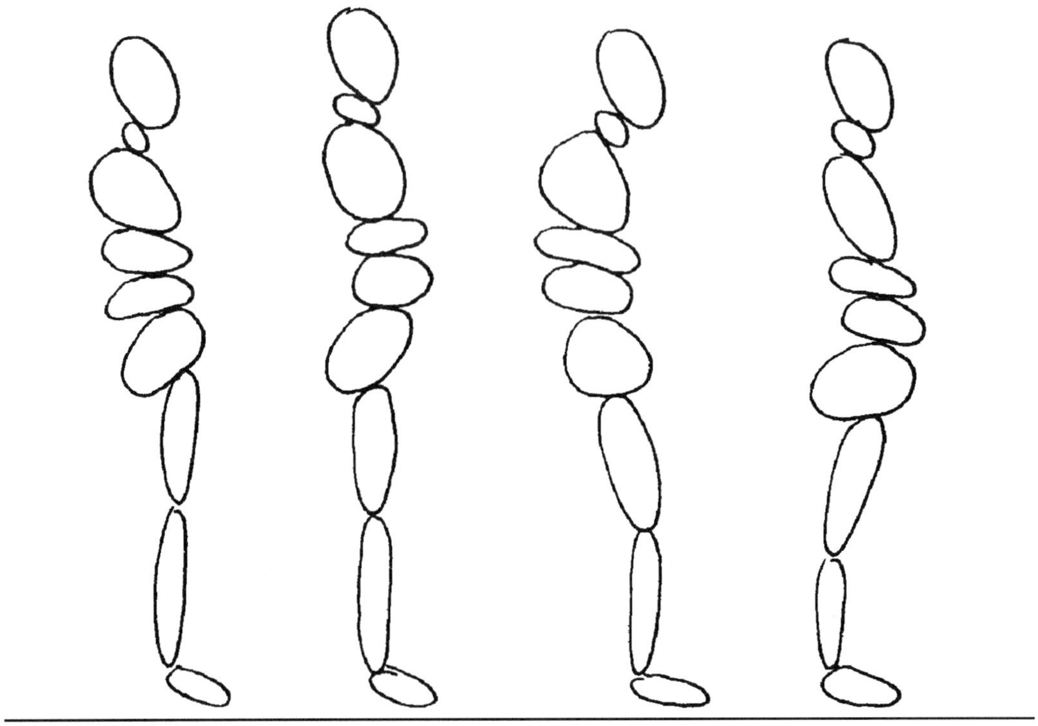

At this point, we can observe the relationship of a segment to its adjacent segment and to its place of most neutral weight bearing. Each of these relationships is important to fully understand the effect of alignment on specific structures in the body. This becomes important when we are deciding what change would be helpful to the overall balance.

Implications

Sometimes we want to observe the relationship between two body segments in order to understand a particular symptom. For example, we could say that perhaps some of their discomfort in the lumbar region is related to the displacement of the pelvis and the lower ribs. Because the lower ribs are not directly aligned over the pelvis, we can look at the force of gravity, or compression, imposed by the ribs onto the pelvis.

Exercise 2.13

On this figure, use a piece of dental floss and anchor it at the front of the ankle, pulling tautly upward in a vertical line, perpendicular to the floor line.

1. Identify the line of gravity (plumb linc).
2. Check the overall segmental relationships to the line of gravity.
3. Observe the spatial relationships of segment to segment. How far forward (anterior) or behind (posterior) is one segment to the one above or below it?

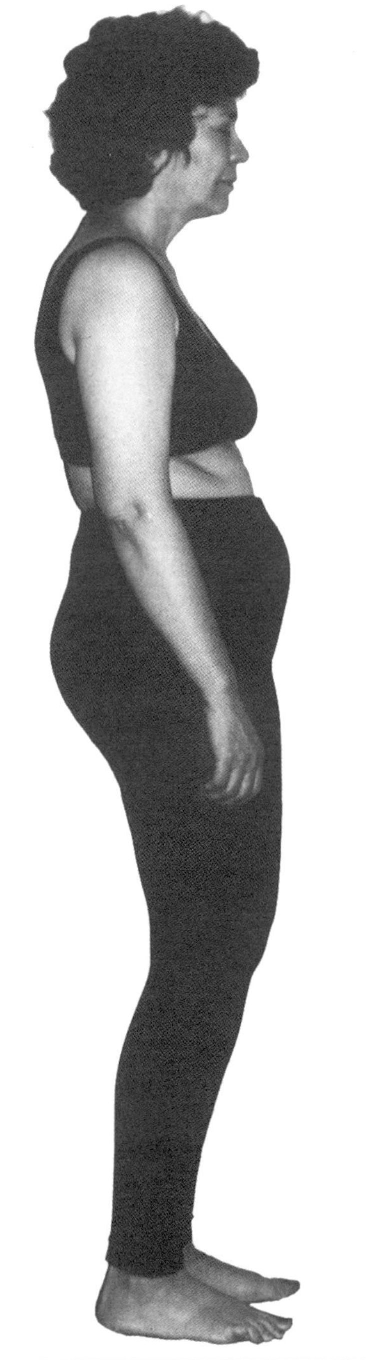

It is useful to relate the position of any body segment to the plumb line in order to determine the change that will best serve the overall balance.

In the example below, would it be best to reduce the anterior-to-posterior displacement of ribs to pelvis by shifting the pelvis backward or shifting the ribs forward—or a little bit of both? This is a hypothetical example. For reasons that will become clearer later, it is rarely useful or realistic to make a singular, one-dimensional correction in alignment.

Demo

In this photo, if the woman were instructed to bring her upper ribs more anterior (forward), what might this do to the rest of her body?

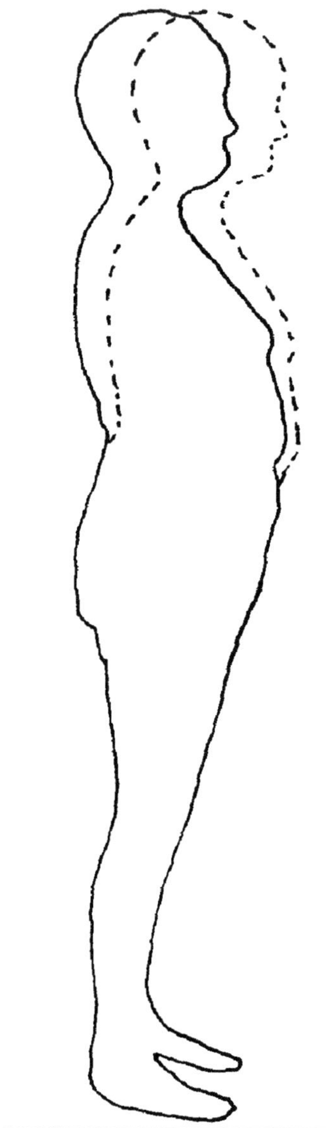

Exercise 2.14

First, place a transparency over this photo. Draw the outline of the body segments, including:

head

neck

upper chest

lower chest

lumbar/abdomen

pelvis

upper leg

lower leg

foot

Remove the transparency and:

1. Mark the geometric centers.
2. Connect the centers.
3. Draw the plumb line perpendicular to the floor from slightly anterior to the lateral malleolus (the outside border of the right ankle).

Now, place the transparency next to the photo. In the beginning you sometimes will see more from the abstract drawing than from the photo. Observe the location of the geometric centers in the body. See which body segments are aligned more centrally. Which is the farthest from the center? Consider the effect of gravity on this system. What effect might the upper segments (head, neck, and chest) be having on the parts below? Are the lower segments (feet, legs, and pelvis) offering direct support to those above them?

Observing these markings, you can begin to speculate about ways that some body segments may be compensating for the misalignment of others.

Exercise 2.15

1. Place a transparency over the photos that follow.
2. Observe these photos.
3. Without drawing body segments, draw a plumb line.

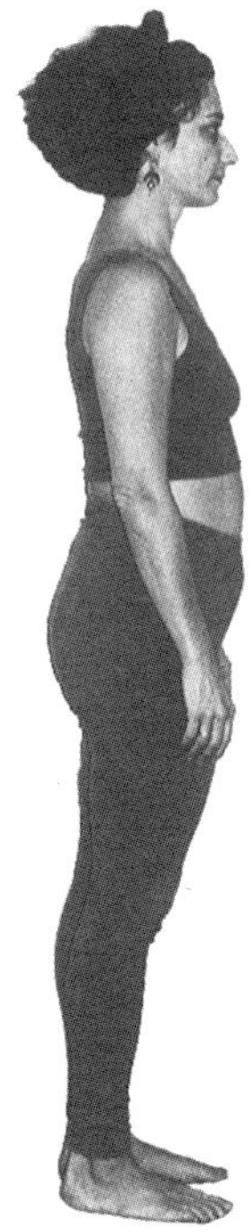

Using your *seeing* skills, without a transparency or dental-floss plumb line, what do you see?

Notation of Placement

Symbols

It is useful to transfer your observations and assessments to the patient's chart without the time-consuming narrative style. These notations can assist your problem-solving during a session. They serve to remind you to compare changes from session to session. They also are useful in writing reports and communicating to your client and other practitioners. In order to make this information easily accessible and visually clear for you and other therapists, we will look at several methods of notating. You will discover what is most useful to you. The following figures show a simple way to quickly notate intersegmental relationships from this perspective.

Arrows can be used to denote the direction of displacement. Begin by placing an X to designate a reference point, which can be either at the top or the bottom of a segment.

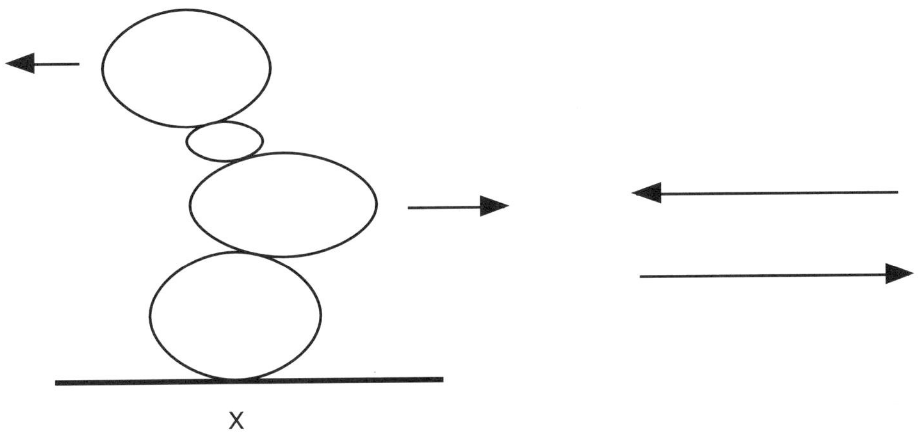

The length of the arrows designates the amount of segmental displacement in relation to the segment immediately above or below it.

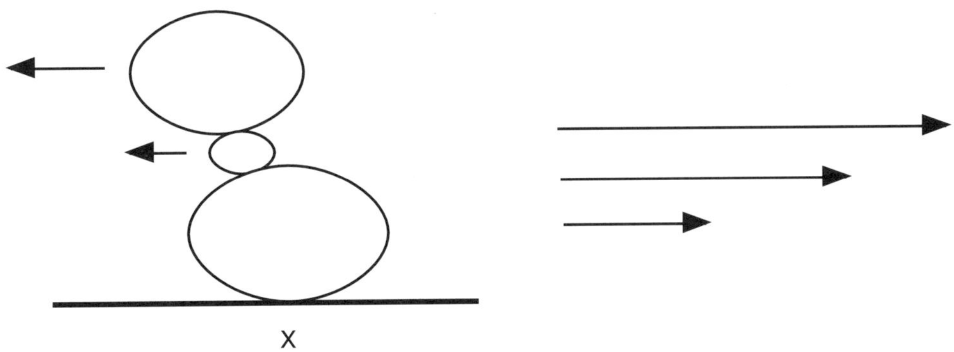

Exercise 2.16

In relationship to the starting point X, which is No. 1:

1. No. 2 is placed in front of No. 1
2. No. 3 is centered over No. 2
3. No. 4 is behind No. 3

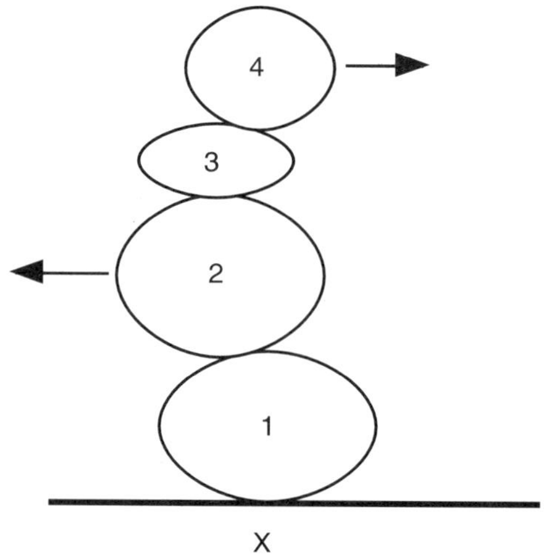

X is the reference point whether you start from above, as shown here.

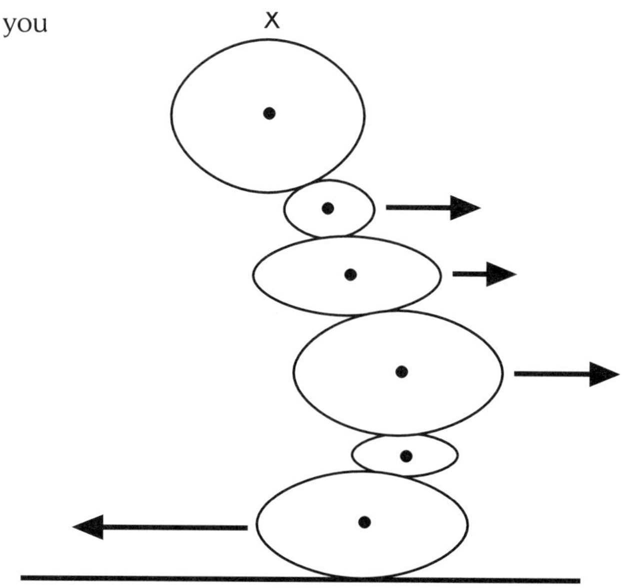

Or below, as illustrated in this figure.

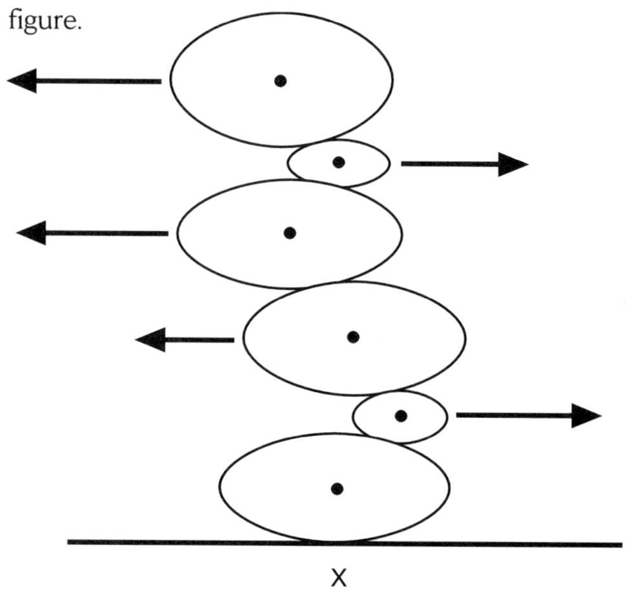

Exercise 2.17

Observe the following sketches of body segments, then notate the positional relationships.

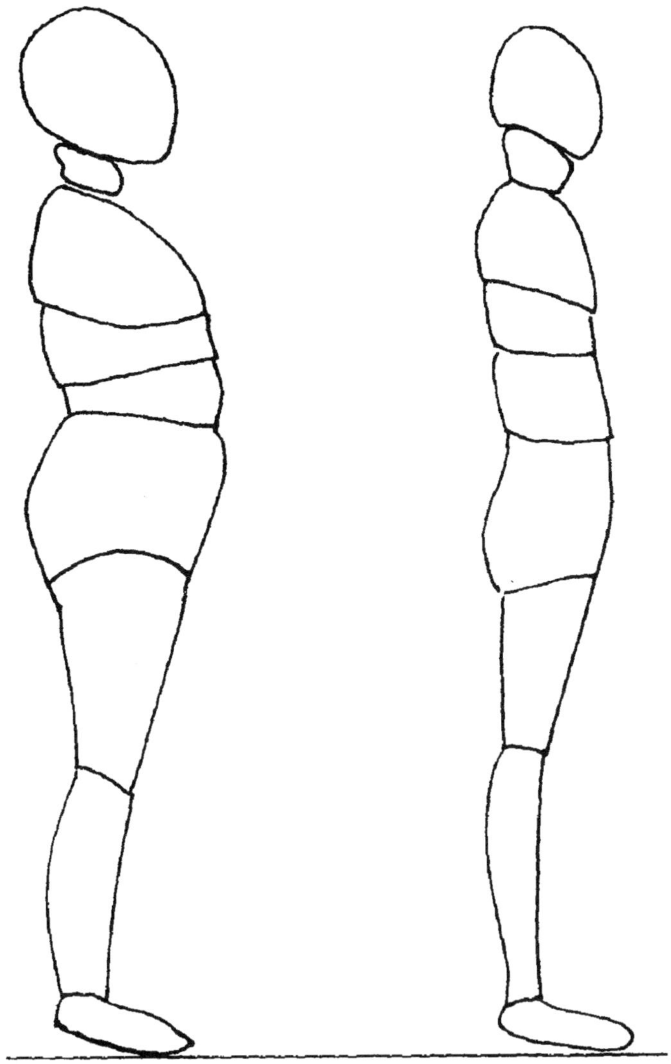

Exercise 2.18

1. Outline the segments.
2. Draw the plumb line.
3. Mark X to identify the reference point.
4. Notate the placement with arrows.

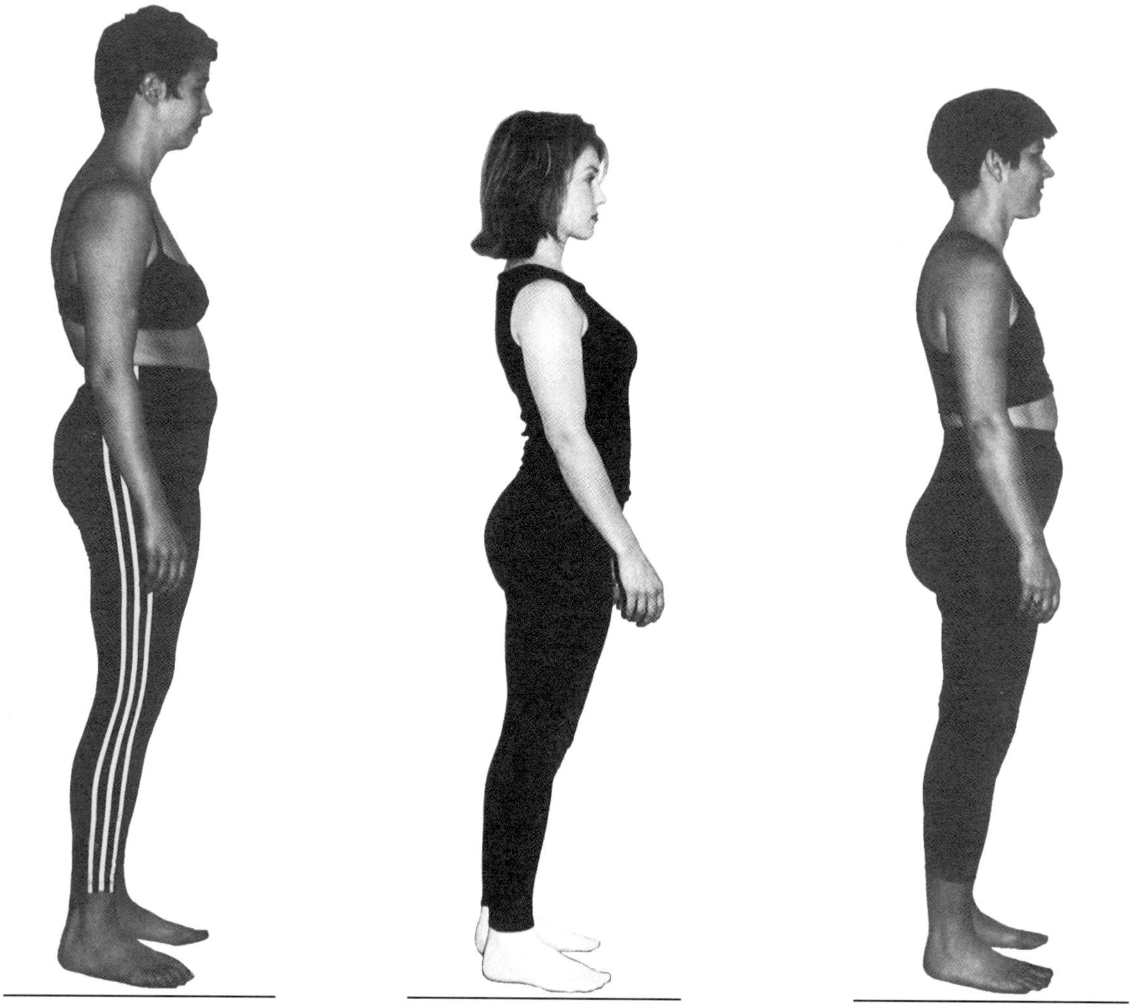

Transferring Notations to the Ball Person Drawing

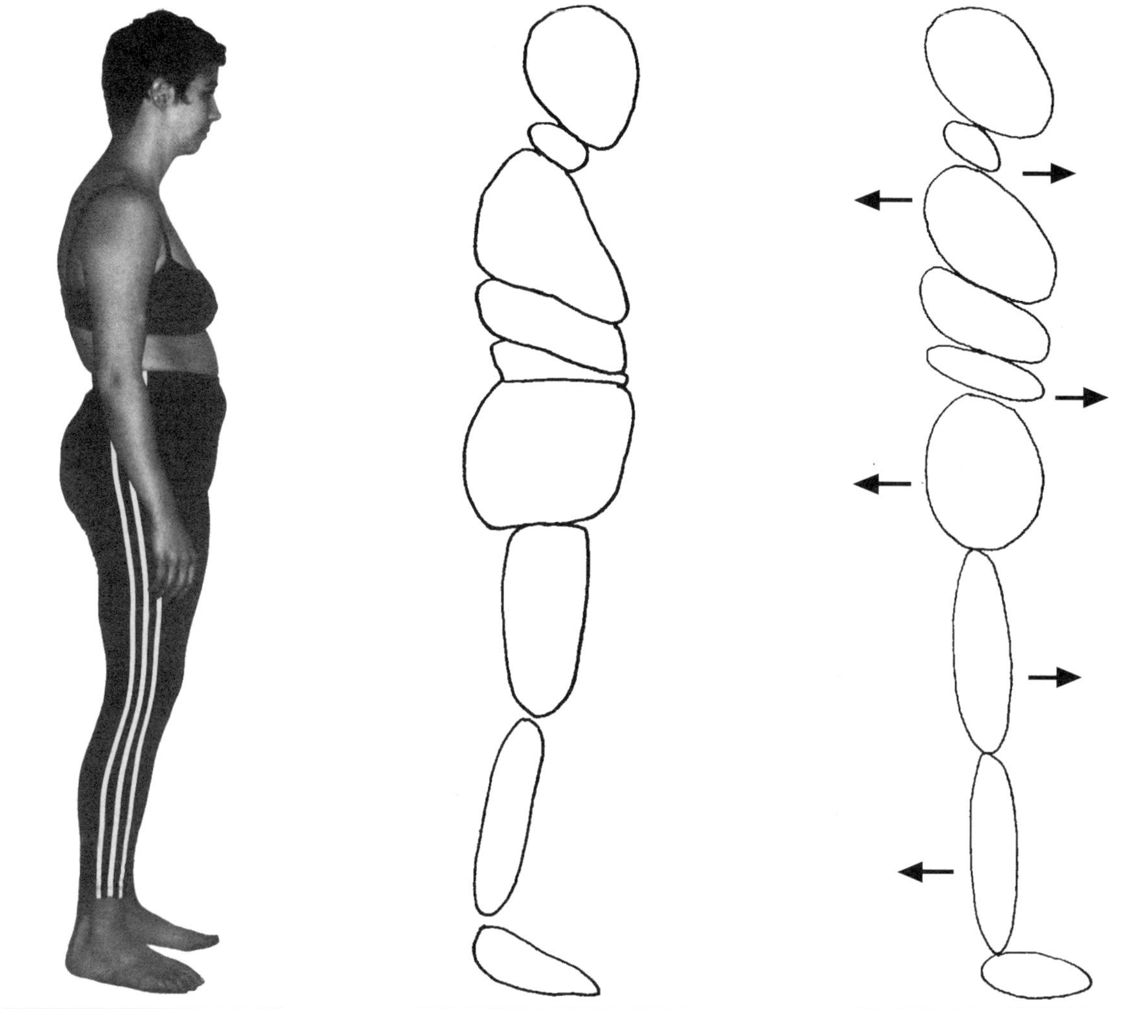

Demo

Below is a demonstration of notating, in the following sequence.

1. Segments are outlined.
2. An X is marked at the starting point to notate placement.
3. The plumb line is drawn.
4. The arrows to notate placement are drawn.

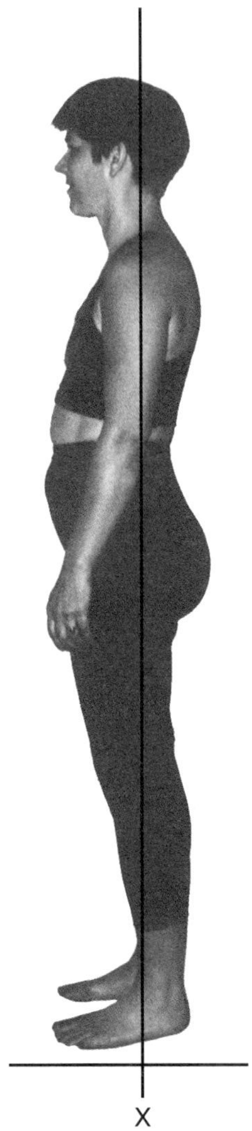

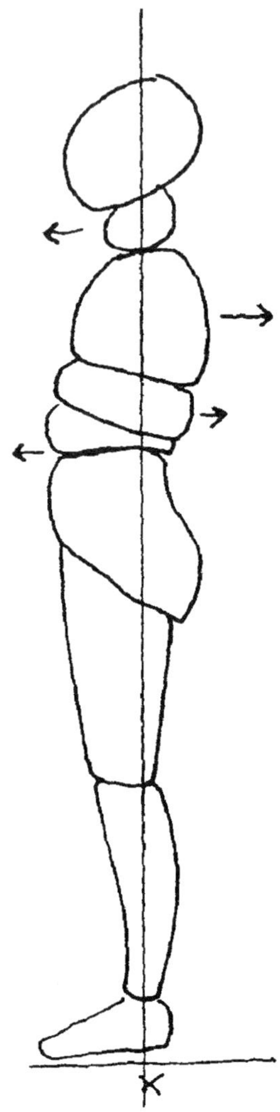

Practice Combining Notations of Alignment

Let's review the concepts and skills we have learned up to this point. Using the following three photos:

1. Outline segment shapes.
2. Mark geometric centers.
3. Mark an X to designate the starting point at the feet or head.
4. Connect the geometric centers.
5. Compare to the plumb line.
6. Notate the direction and amount of anterior-to-posterior displacement with arrows.
7. Based on the information in numbers 1–6, where is the major point of weight bearing on the foot (heel, forefoot)? Place an arrow to designate the greatest weight-bearing area on the foot.

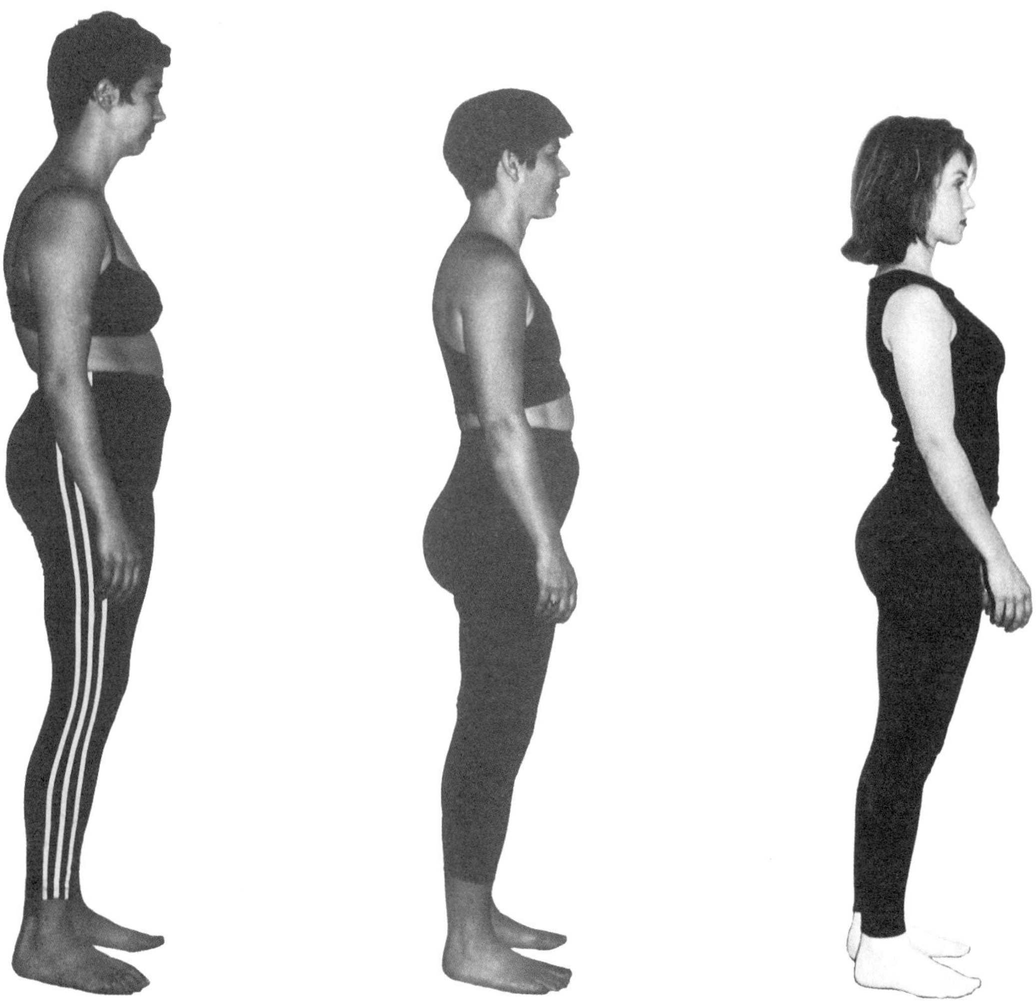

Transferring Notations from Photos to Standard Body Charts

Some therapists use photographs regularly to track changes in a client's posture. These can serve as a useful measure of progress and a visual reference for study. The notations with which we have just worked can be useful if they are marked directly on a transparency over the photographs.

Whether you are using photos or notating the person while he or she is in the standing position, it is useful to transfer your observations to a body chart. This can be done by putting the same notations of placement onto the chart as follows.

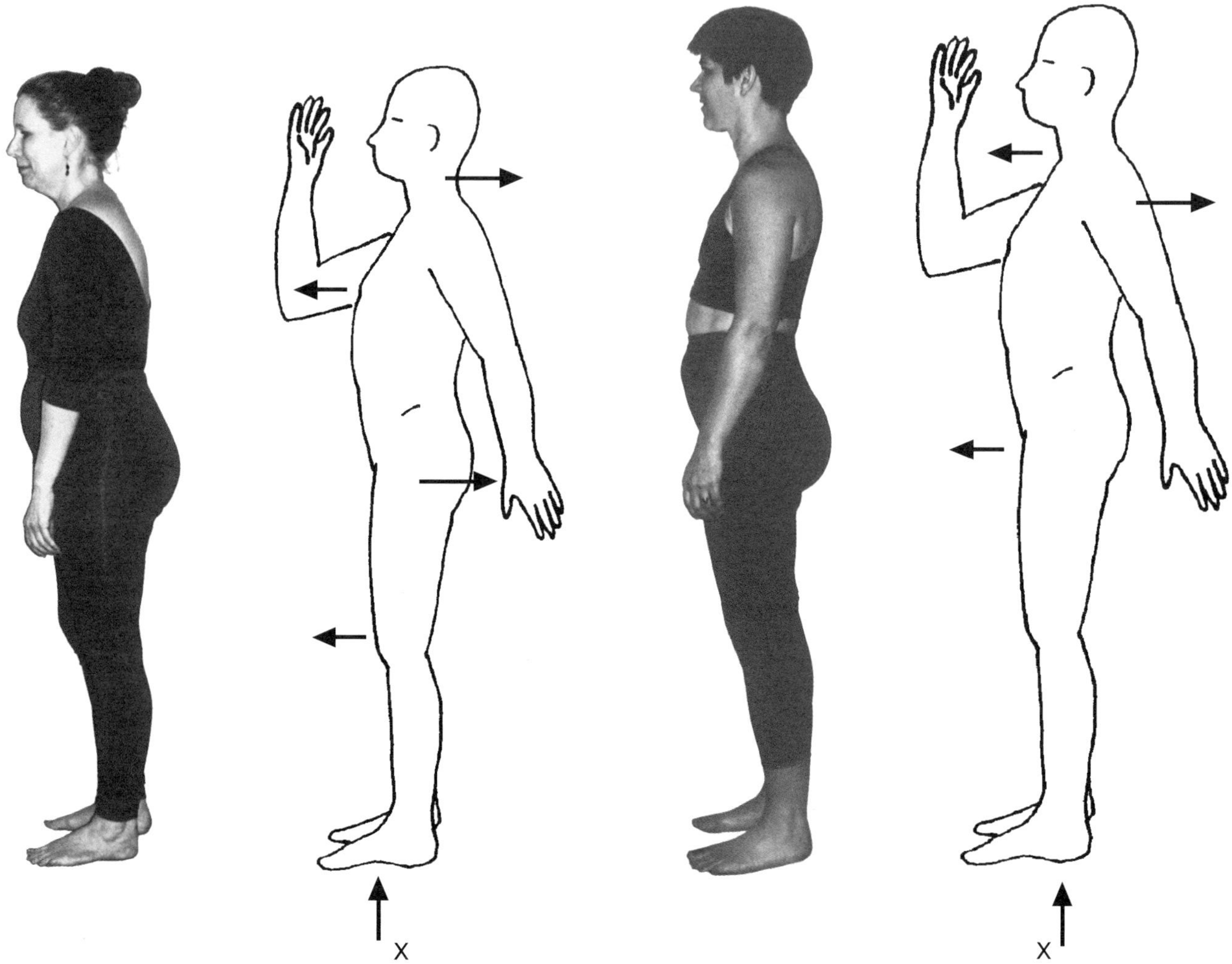

Practice

Notice that in the previous examples, the use of longer and shorter arrows helps to keep clear the relationship between one segment and another, as well as the relationships to the plumb line.

Look at the following photos and compare each to its abstract drawing, which also demonstrates segmental body placement.

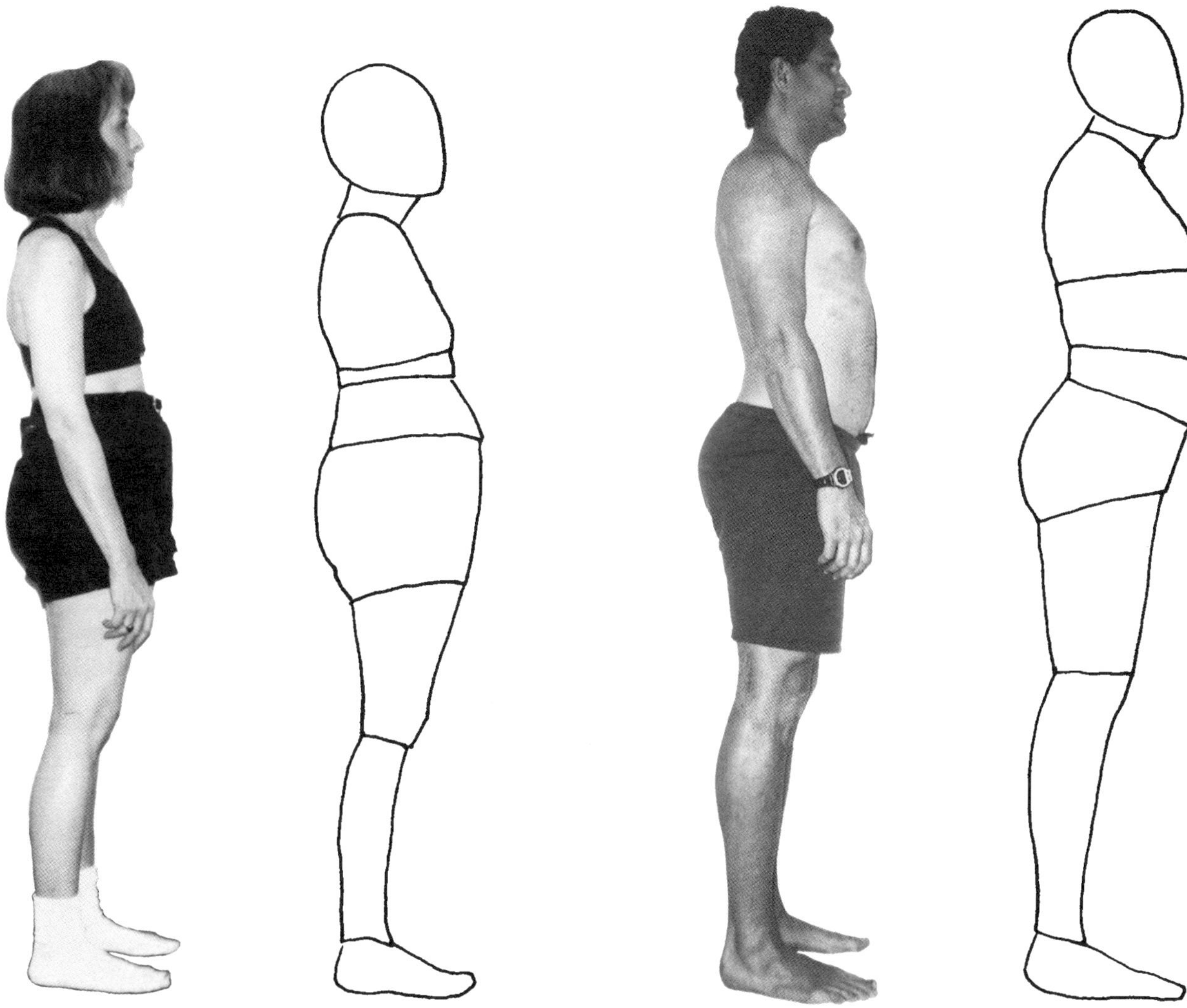

Notate the alignment in the following photos onto the ball person drawing.

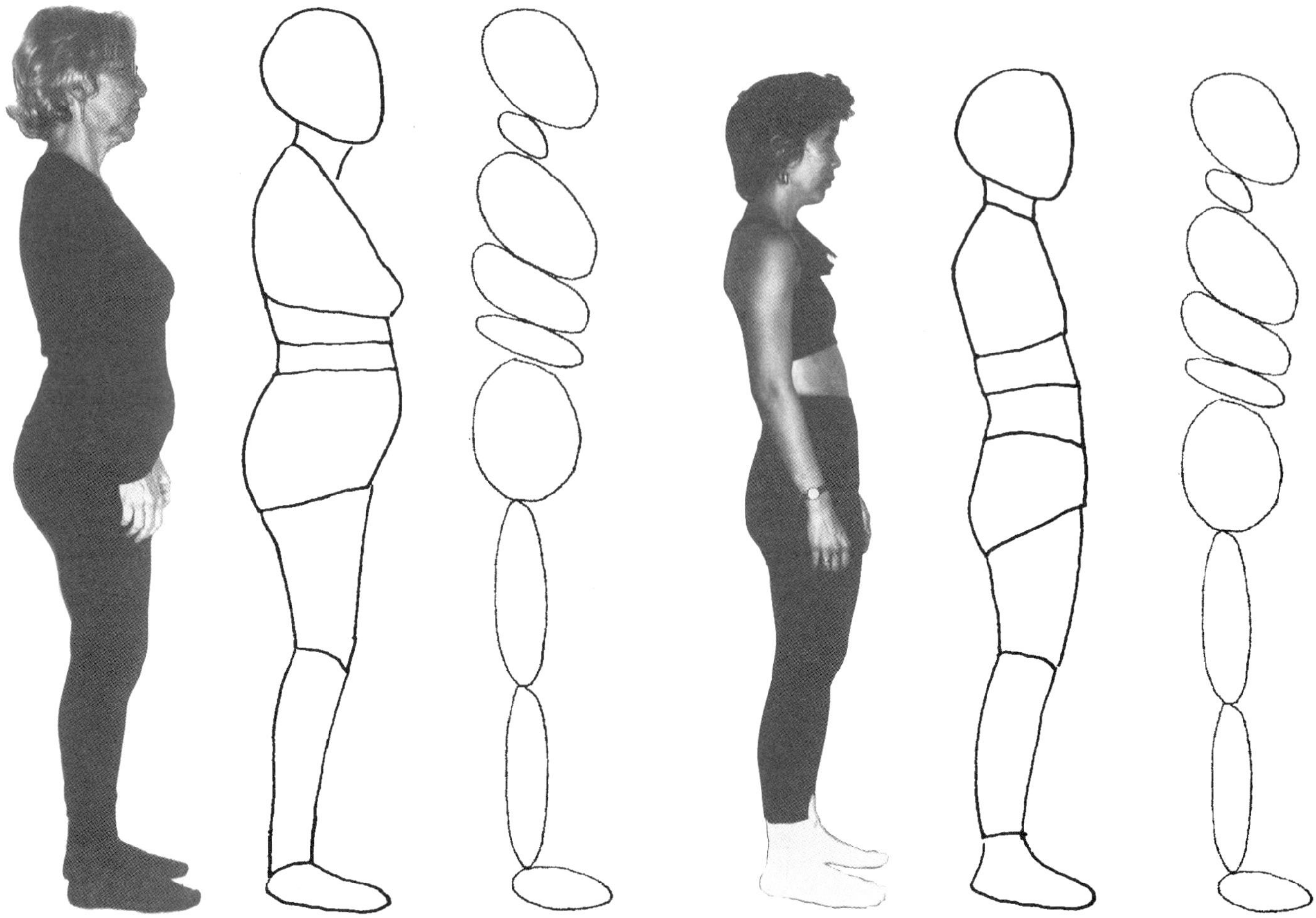

After you have made your marks, cover the photo and abstract body drawing and look at your notation. Visualize the alignment of the body you have notated. Uncover the photo and see how well your notes have enabled you to picture this client. This would be a good time for you to see if you can approximate the anterior/posterior relationship by using your notations to put this pattern into your own body. Note which markings were missing or misleading. Alter your notation so that you can reproduce the alignment accurately without the photo. Refining your system of notation will enable you to compare accurately the alignment of a client from one session to the next and to document progress over time.

Standard Body Charts

These are drawings that include realistic body contours and landmarks. Many therapists use these to have clients mark areas of pain or injury. You also may want to use them to record your findings. See the following examples.

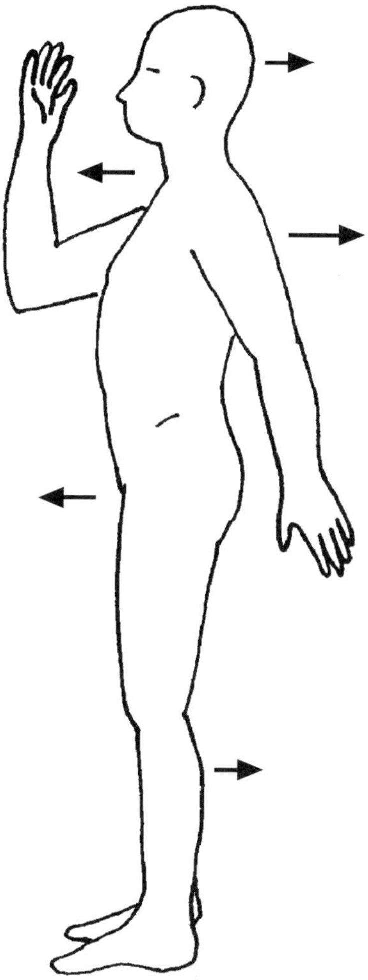

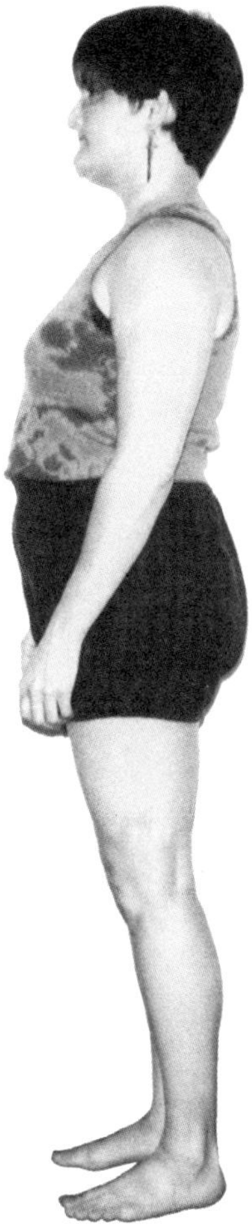

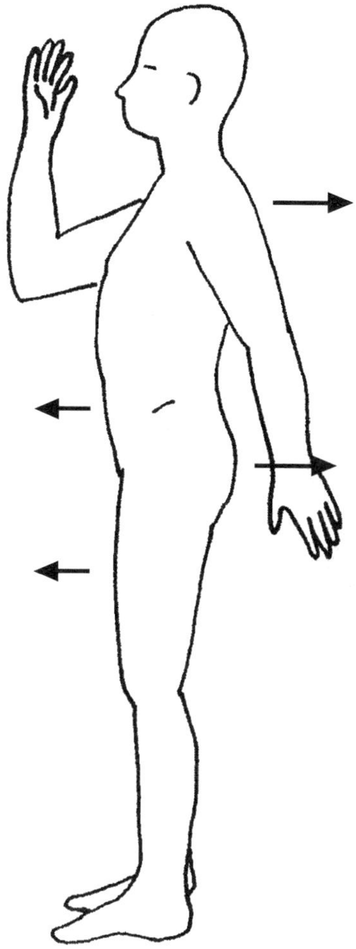

Exercise 2.19

Now use these body charts to note the alignment in the following photographs.

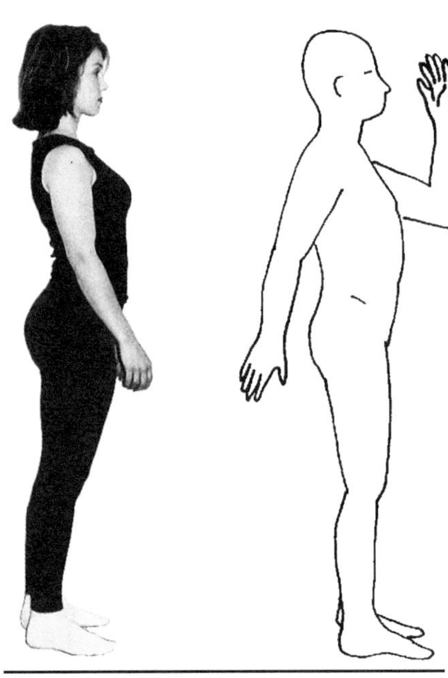

A

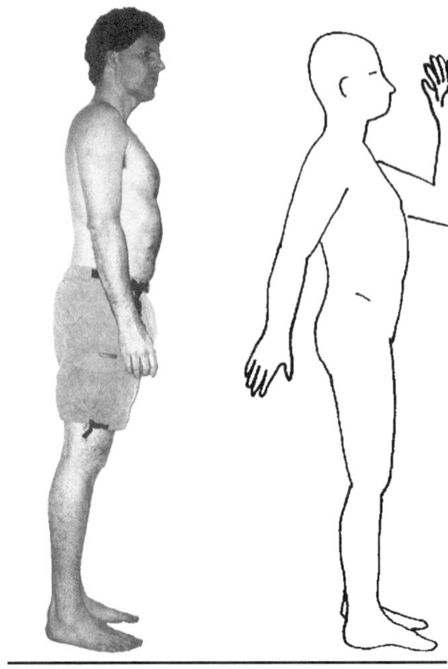

B

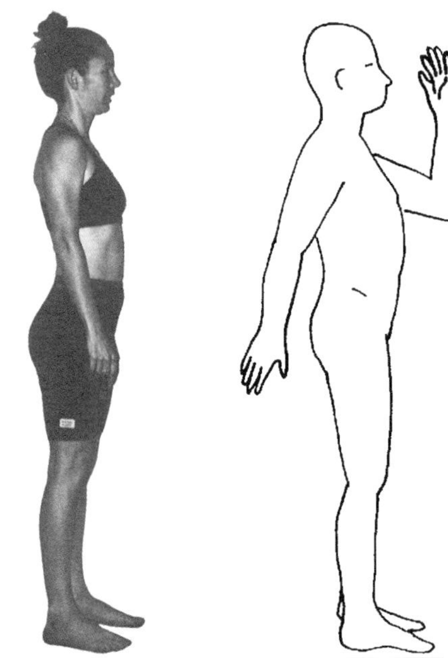

C

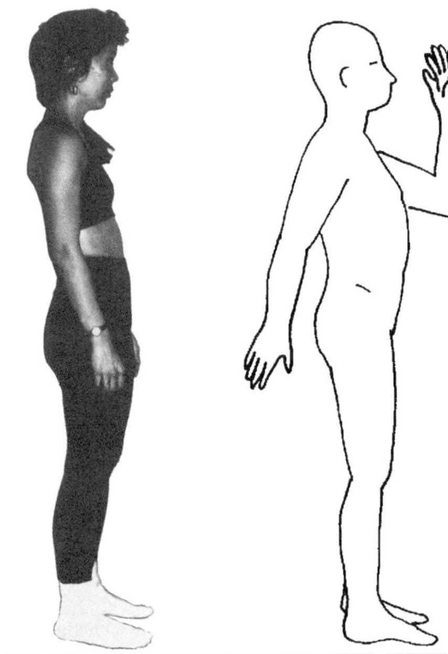

D

Preview of Other Notation Symbols

As we continue, we will expand our group of notations by adding different symbols for various segmental placements. A preview of those symbols follows.

Notations

Placement

right and left*

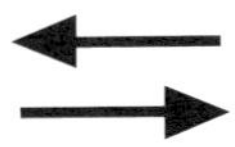

*Looking at a photo is a mirror image—arrows are opposite.

up and down

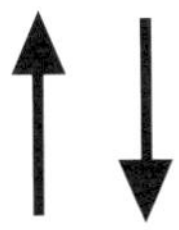

anterior and posterior

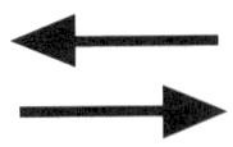

Weight-Bearing Center

Tilts

to mark up-to-down angle

Rotations

to indicate one side forward and one side behind

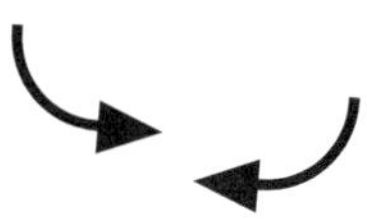

or many segments rotated in same direction

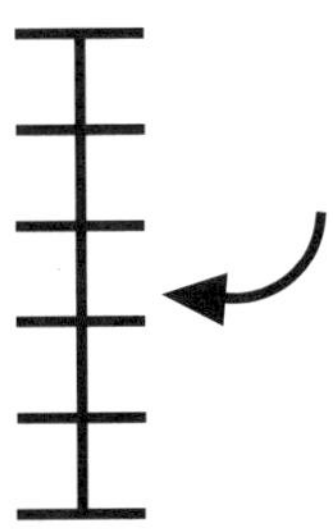

Anterior/Posterior Rotations

Up to this point, we have been emphasizing one plane of alignment at a time. Within each plane, it is important to view the body from both sides in order to get a total picture. Consider the two side views of the bodies below. Mark the centers of mass, connect the centers, and mark the plumb lines. What do you see?

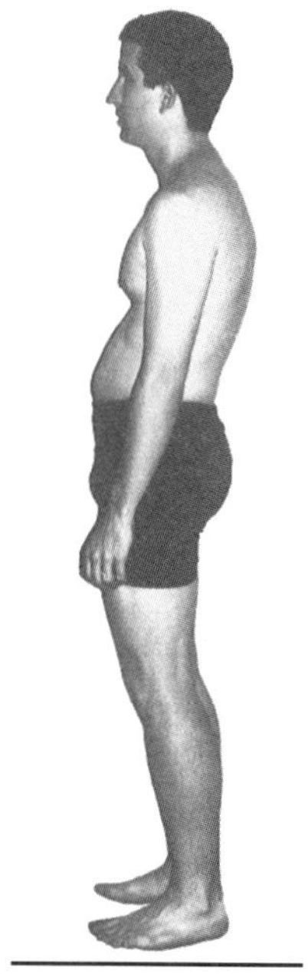

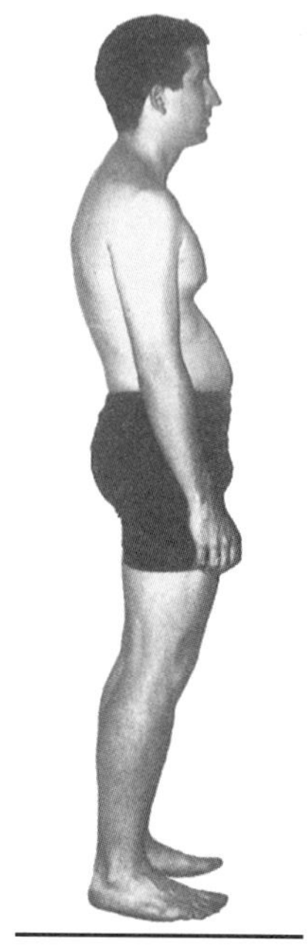

In comparing the two sides of the body, we clearly see that often their alignment is not the same. How can this be?

In addition to being displaced, body segments also can become rotated. As we have seen in the previous figures, a rotation can, for example, place one shoulder on the plumb line while the other is considerably posterior. The pelvis can be placed centrally on one side and forward of plumb line on the other.

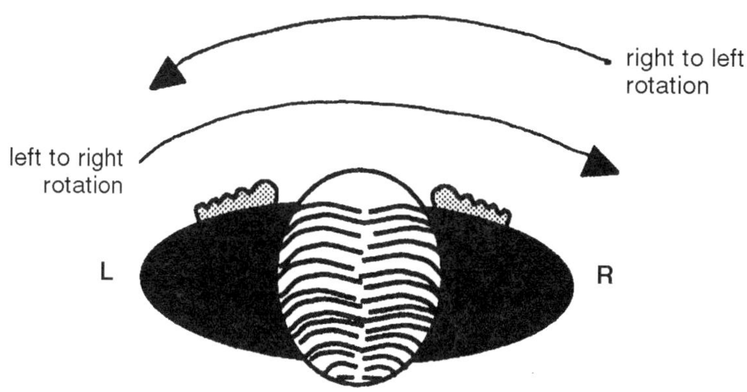

We describe rotations as occurring from right to left (which means the right side rotates toward the left side), or left to right (left side rotates toward right). I am including the front view photographs to show that rotations also can be recognized from the front or back. Observe the rotations in the following photographs.

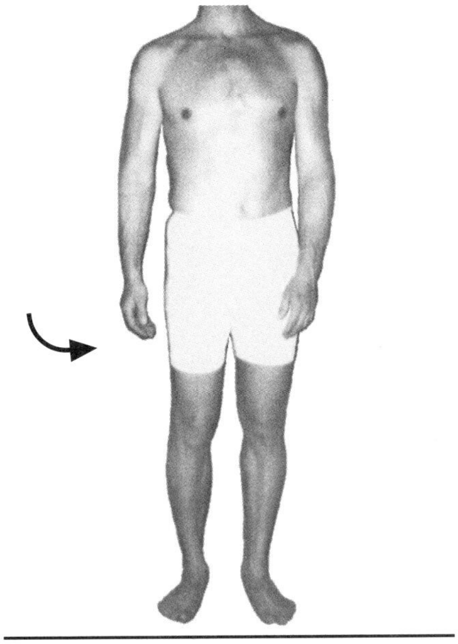

Right to left rotation of pelvis

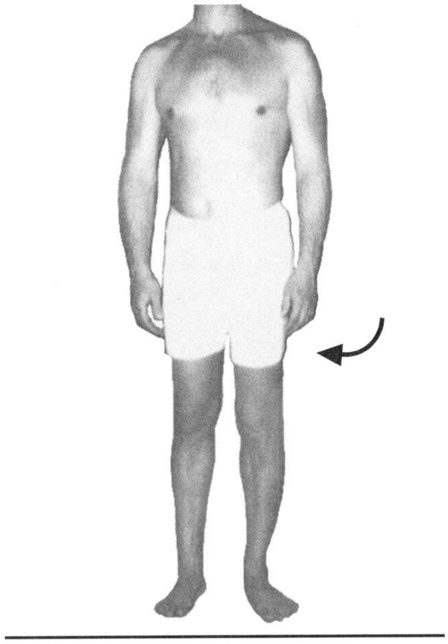

Left to right rotation of pelvis

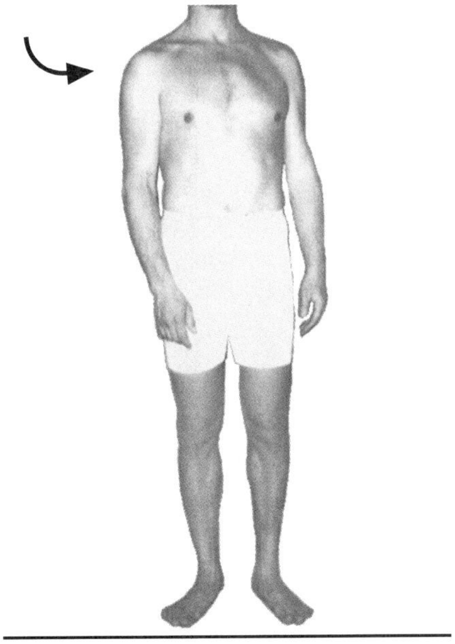

Right to left rotation of chest

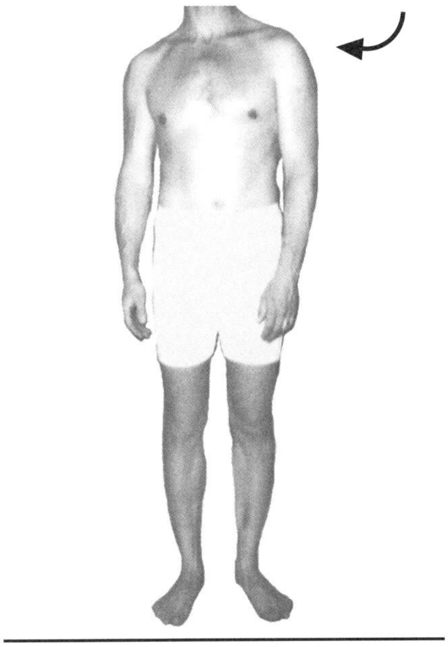

Left to right rotation of chest

Counterrotations

We begin to see that a body generally does not have one part in rotation without also having other rotations. Just as a particular pattern often includes displacement forward and backward of the plumb line, it often has rotations in one direction and the opposite direction. We will call these counterrotations.

Let's observe one more frontal view before going back to the sagittal plane. Note the possible combinations in the following photos.

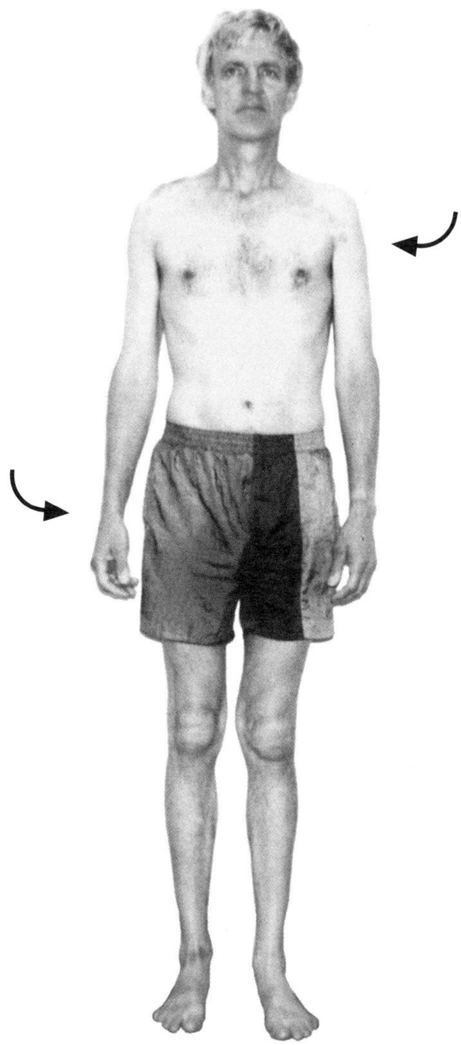

These are examples of some rotations. As you will see later, there are many more possible combinations of rotations and counterrotations. We will look at compensations in more detail in Chapter 3 and see the role of each of these rotations in creating overall balance.

Exercise 2.20

Draw geometric centers on the following photos and determine whether there is simple displacement or rotation for each segment by comparing the two views. Note that if the amount of displacement of a segment is the same in both views, this indicates that the whole segment is shifted forward or backward. For now, if both sides are not in the same relation to the plumb line, a rotation is indicated.

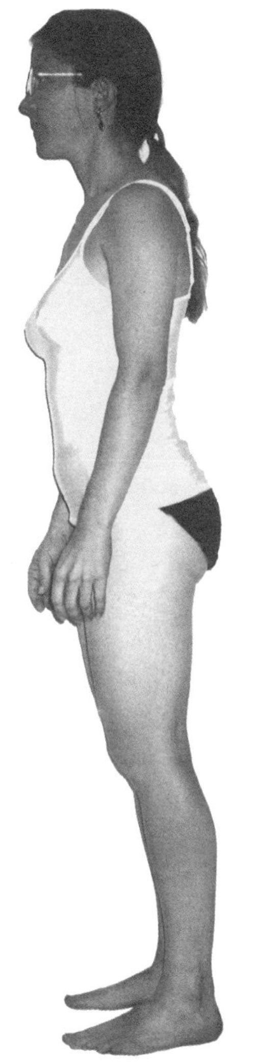

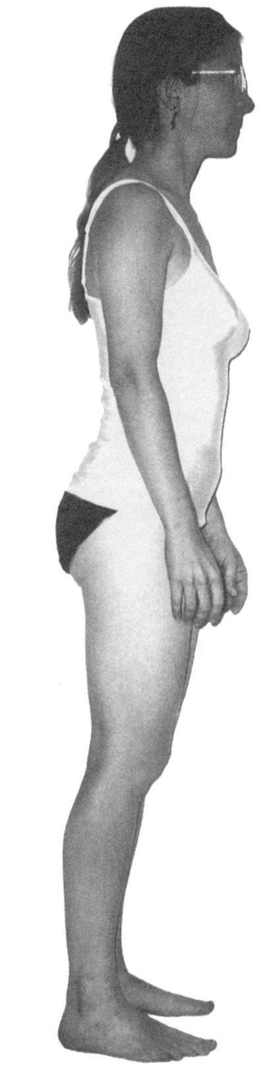

Notation of Rotation—Rotations in the Sagittal Plane

In order to document rotations, curved arrows are visually clear and can be used to distinguish rotations from simple displacements. Consider these examples.

The curved arrow forward indicates that segment No.1 is rotated forward or backward in relation to the plumb line.

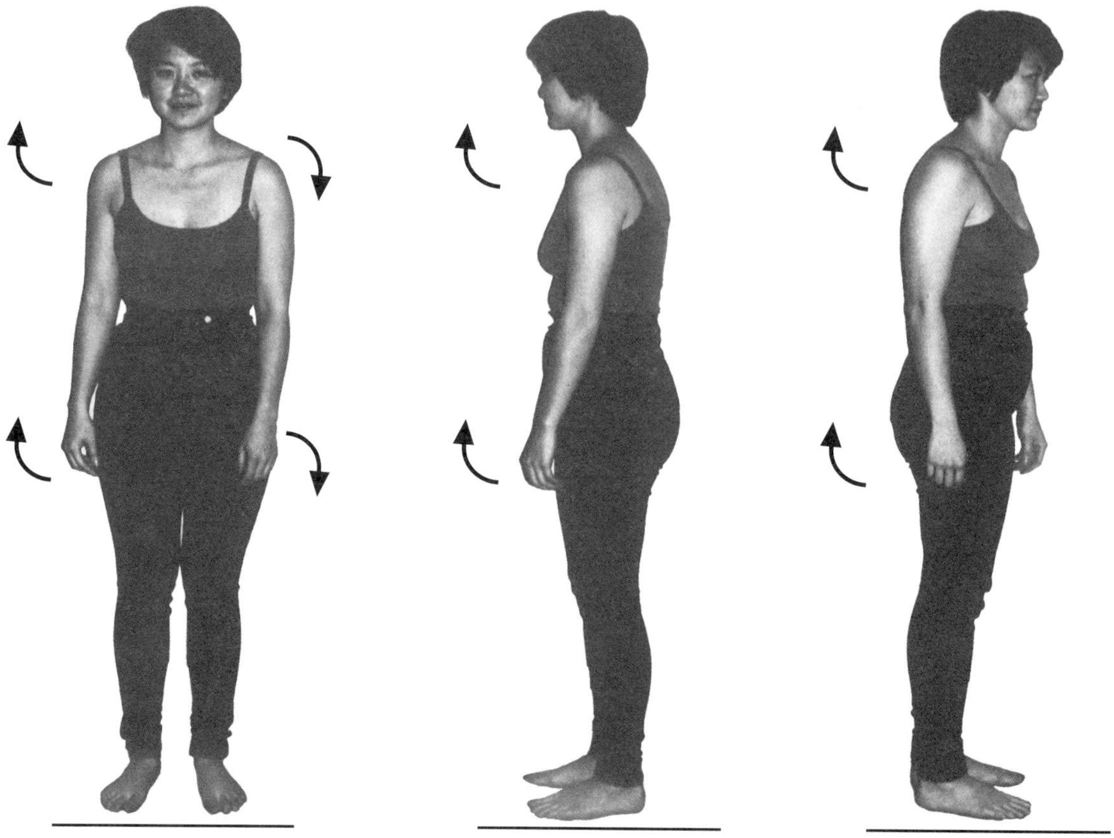

Exercise 2.21

Mark the placement and rotations for the following bodies.

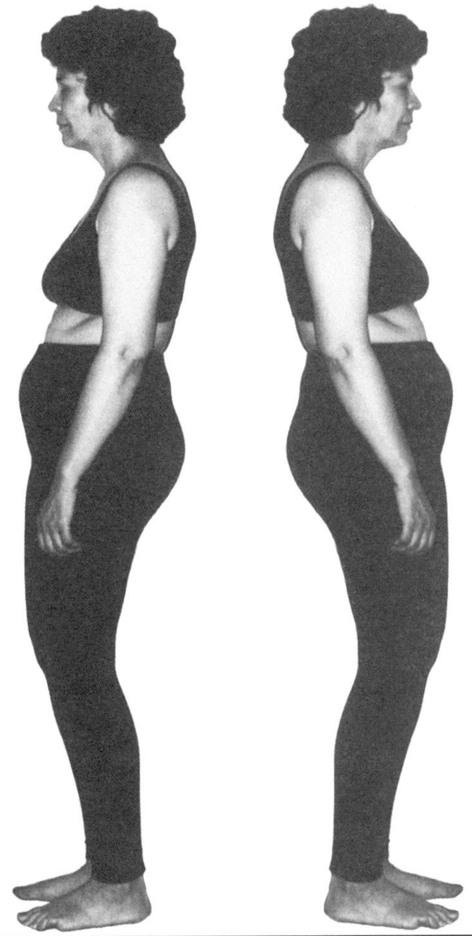

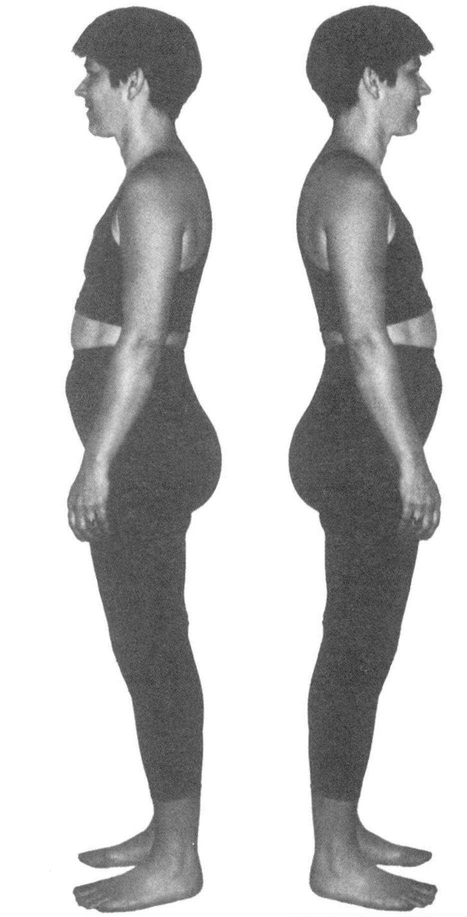

At this point, you should obtain several sets of photos from the front and back views and the right and left sides. You may find it interesting to work with photos of yourself, as well as those of clients and friends. You may wish to tape your photos to this page so that you can continue to use them for reference.

Sagittal Plane

Observe the following photos. Without using a transparency or making notations, observe which side gets your attention first. Then which segment? This segment is in relationship to which other segment? Do you see more displacement or rotation?

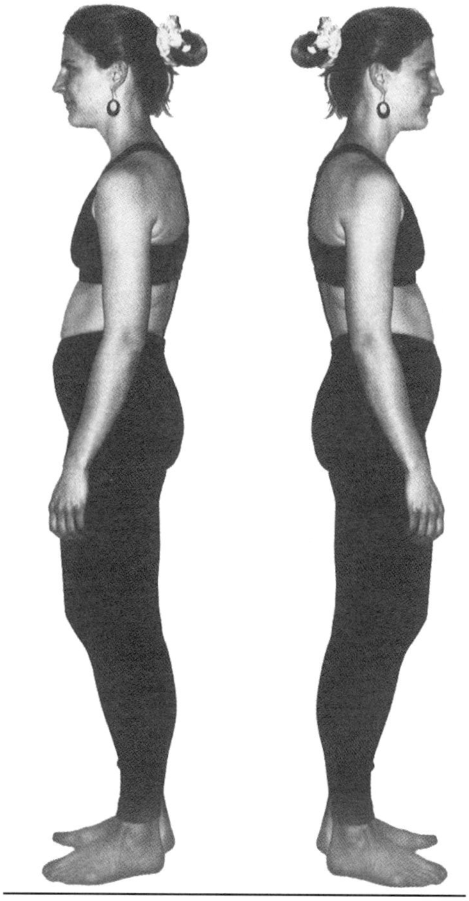

Seeing Alignment in the Frontal Plane

Observe the progression of illustrations to see how the body segments define the ball person.

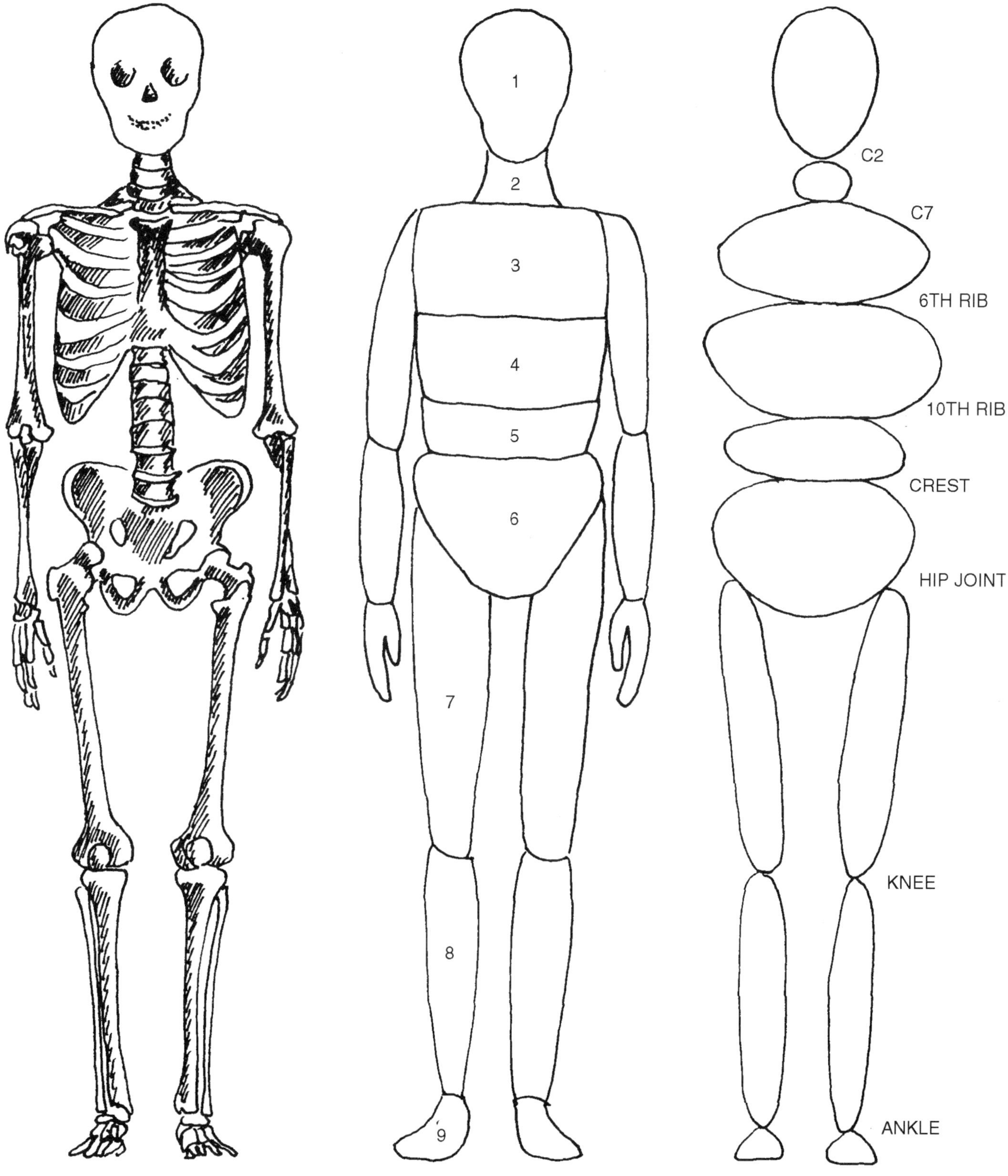

Outline Segments—Front and Back of Body

To break down your assessment more specifically, first take a piece of paper and cover all the body segments below the head. Observe the segment, draw it onto a transparency and then add the segment below. Move sequentially down the body, adding each body segment in the following order:

1. Head
2. Neck
3. Upper Chest
4. Lower Chest
5. Abdomen/Lumbar
6. Pelvis
7. Upper Legs
8. Lower Legs
9. Feet

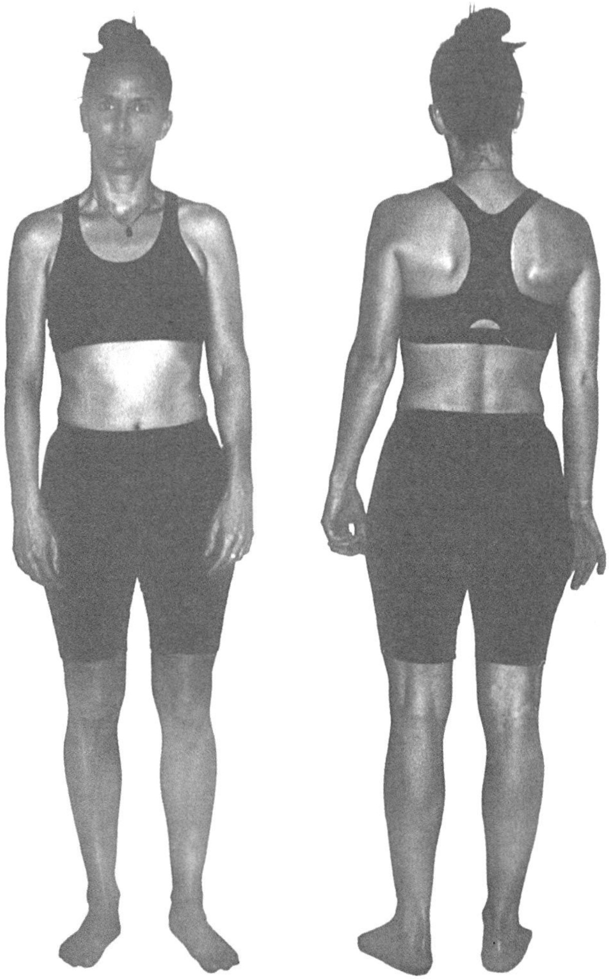

Segmental Centers and Relationship to Placement

You will end up with the following notation.

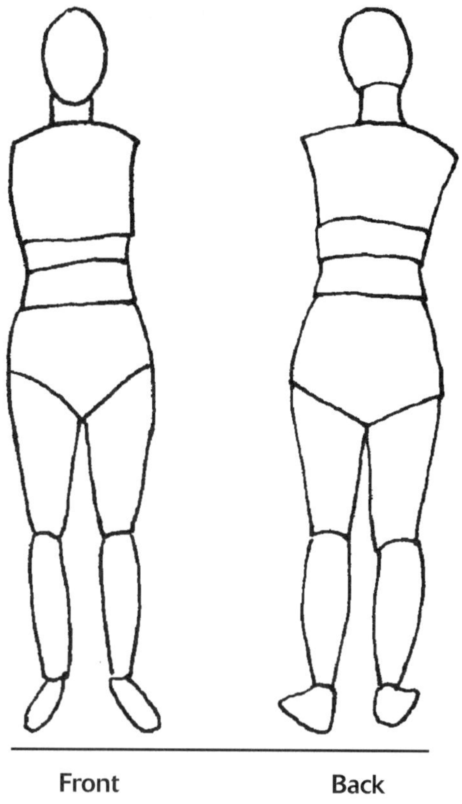

If we add the segmental centers and connect the centers, we will see the following.

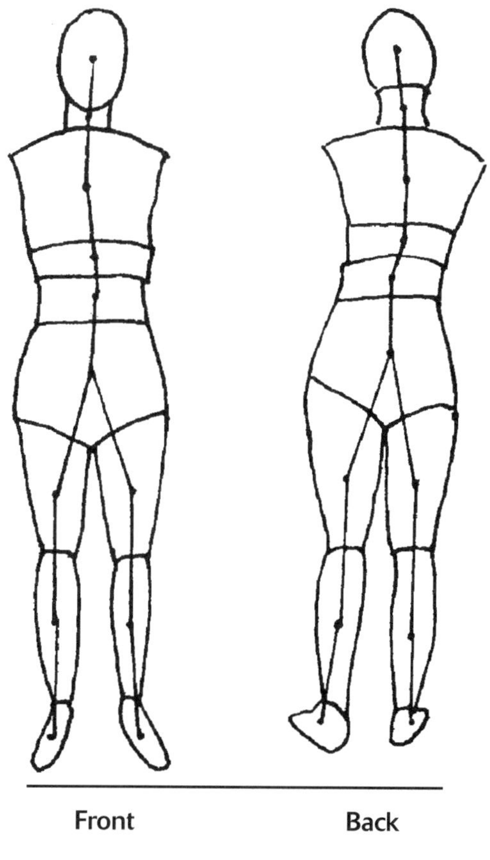

Exercise 2.22

Use the space below to outline the nine body segments in the following two photos. Then add the segmental centers. Next, connect the centers in order to see the positional relationships.

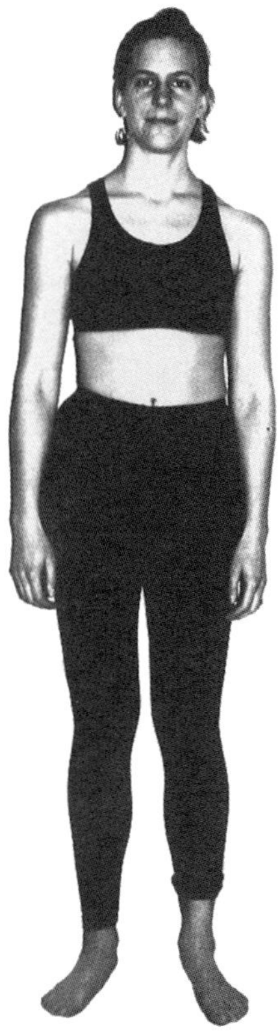

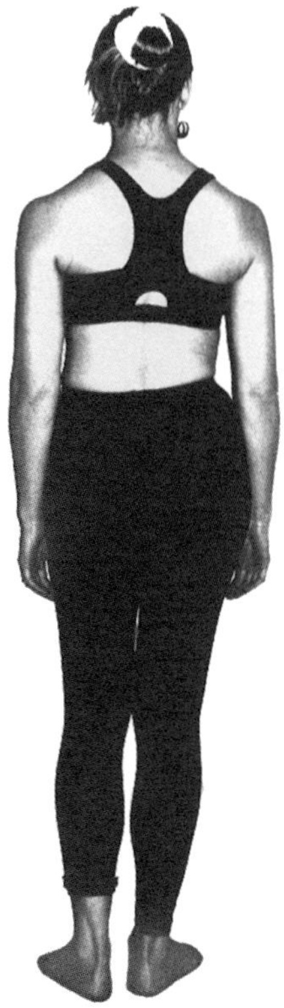

Notation of Right-to-Left Placement in the Frontal Plane

Now, let's observe the right-to-left relationships of the segments to each other in the frontal plane.

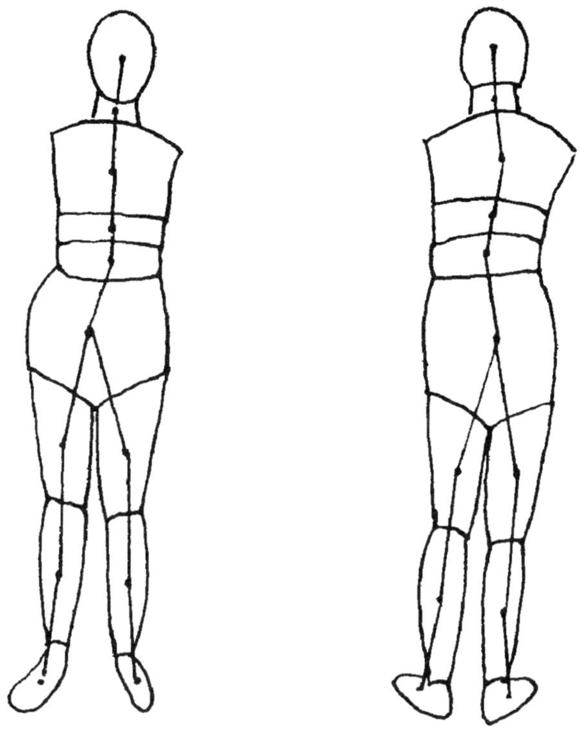

The zigzag line created by connecting centers makes the placement obvious.

We now can progress to documenting placement right or left with arrows. An X marks our reference point.

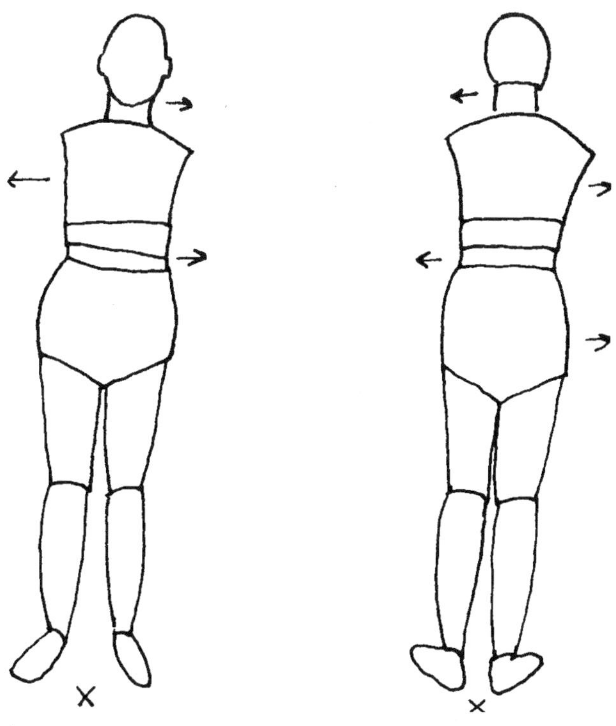

Exercise 2.23

1. Mark the segmental centers.
2. Connect the centers.
3. Notate placement of one segment to the segment above or below it (right-to-left with arrows), starting from the feet and going toward the head.

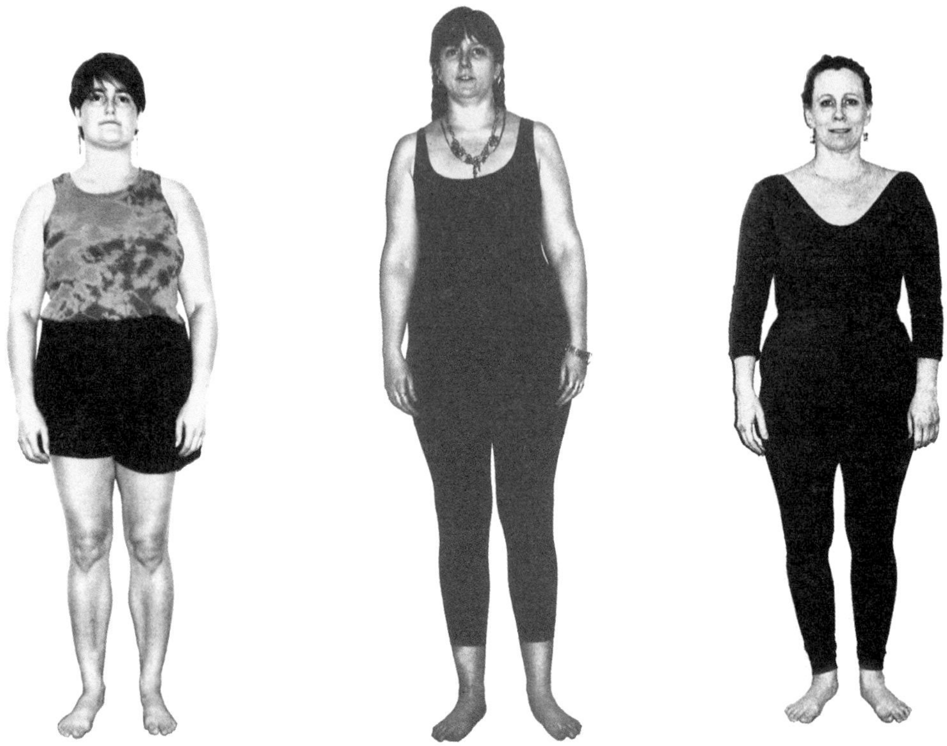

Transfer the X and arrows to the following charts.

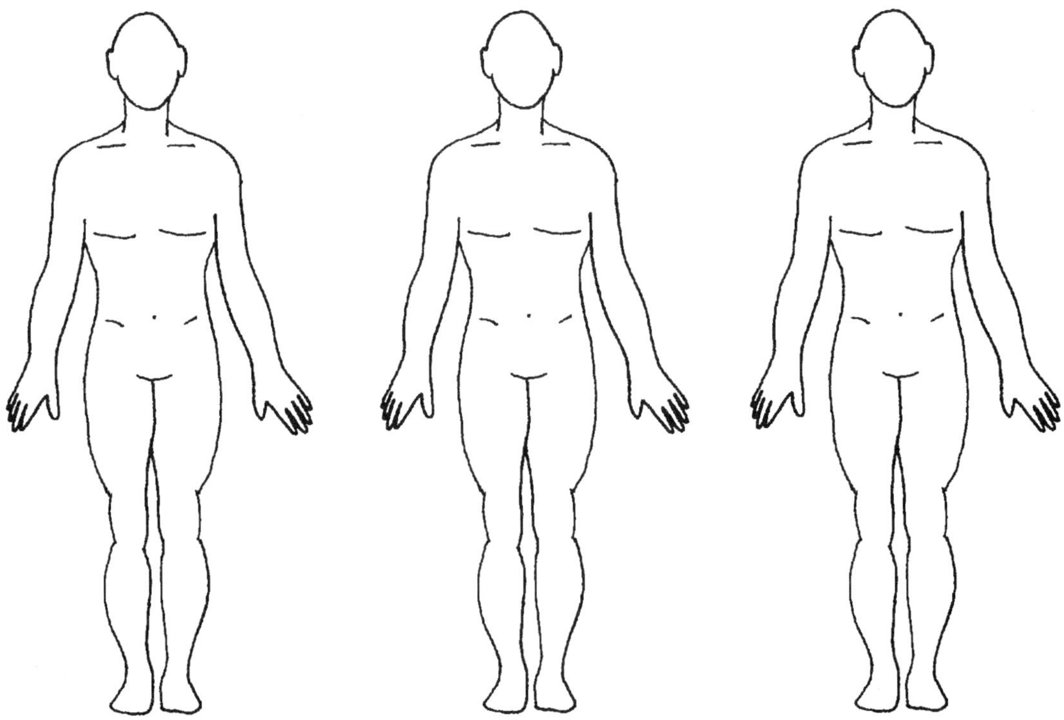

Notation of Gravity and GRF

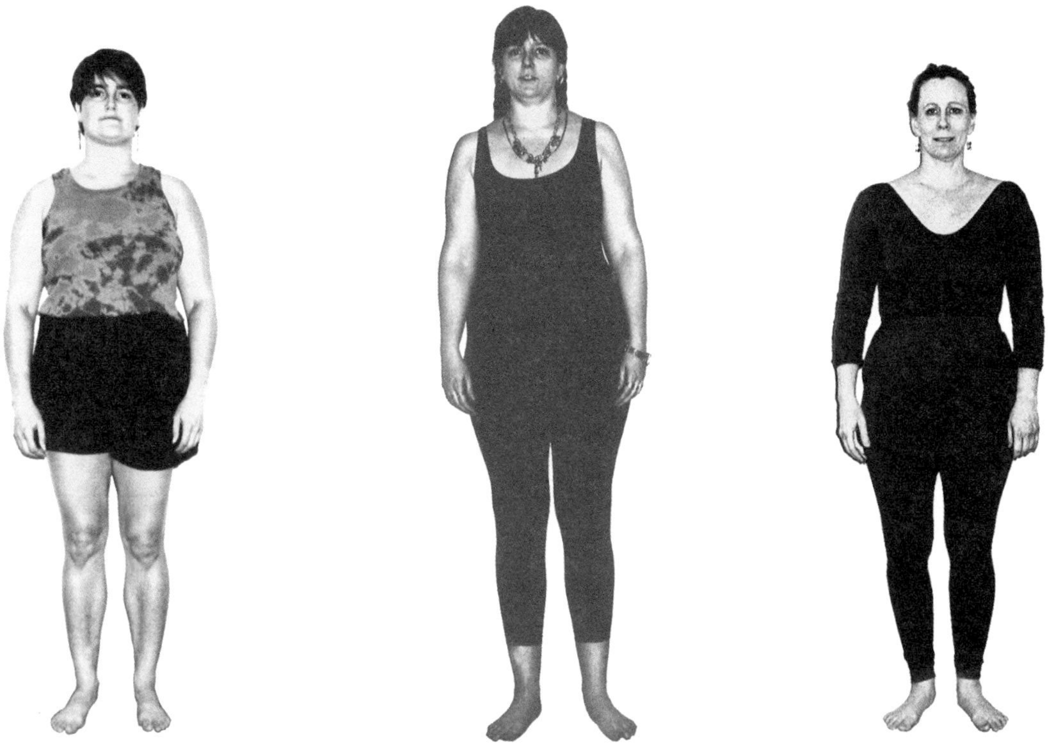

Observe that these forces travel in the same direction as the arrows of right and left placement. The longer the arrow, the greater the force.

The Plumb Line

Refer to p. 38 for a reminder of how to determine an object's center of gravity and mark a plumb line to illustrate the right-left and front-back divisions of the body.

Exercise 2.24

Use dental floss to identify the plumb line in these photos. Notice that you cannot line up all the segments. If you start midway between the feet, both people have their upper body shifted right.

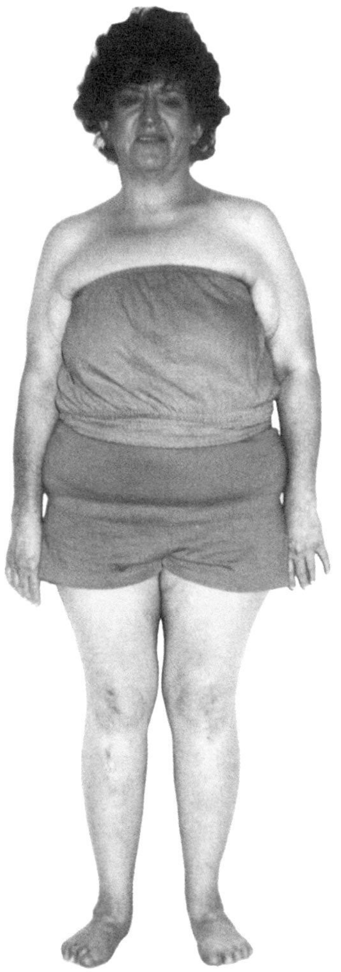

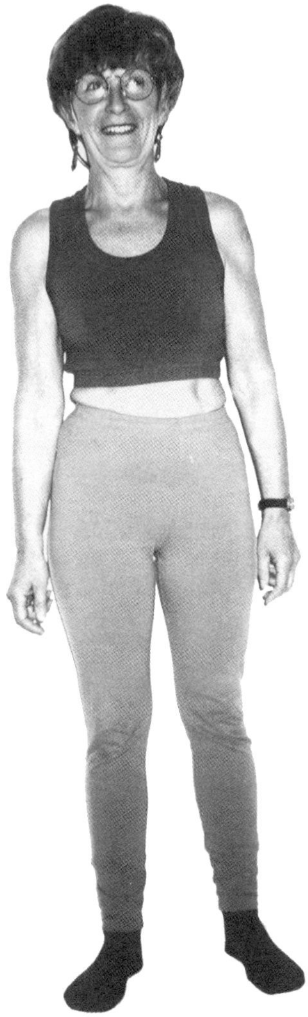

Result: Compared with the standard and familiar model of posture, these bodies are not *ideal,* but they are maintaining the best available posture. They demonstrate common patterns, and you can easily gather the information of right-to-left displacement from this exercise.

On a transparency, by adding the plumb line to the photo of our client, it is easy to see that from the front view, the pelvis is placed to the right of the plumb line, and the mid- and upper trunk are placed to the left. Now, add arrows to designate the placement of segments to the plumb line.

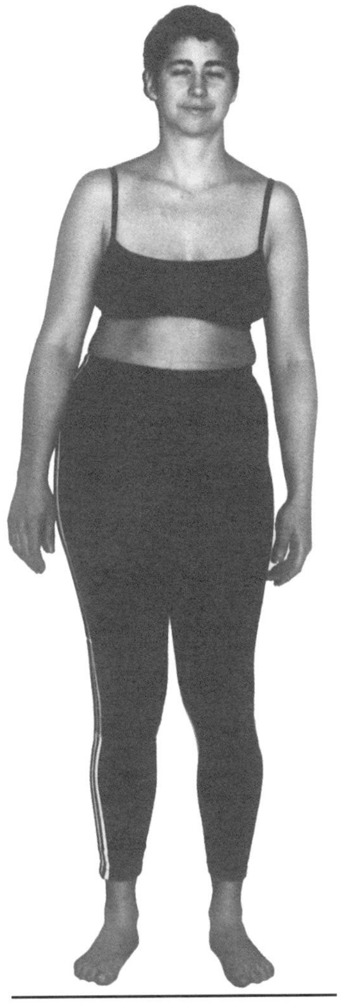

Front

Compare the final notation with the photo.

On a transparency, notate the plumb line and the placement of the segments to the right and left.

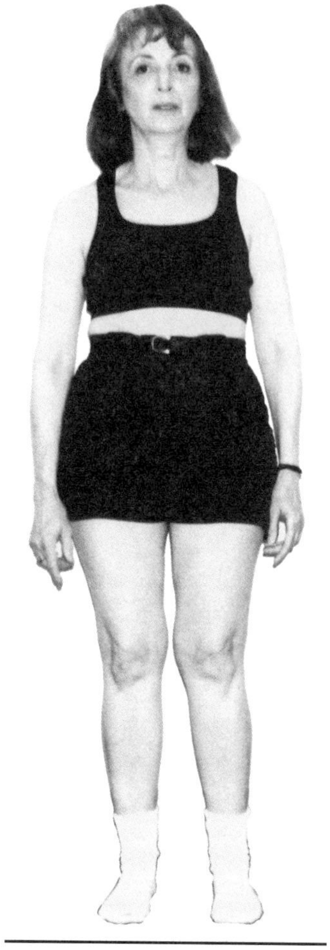

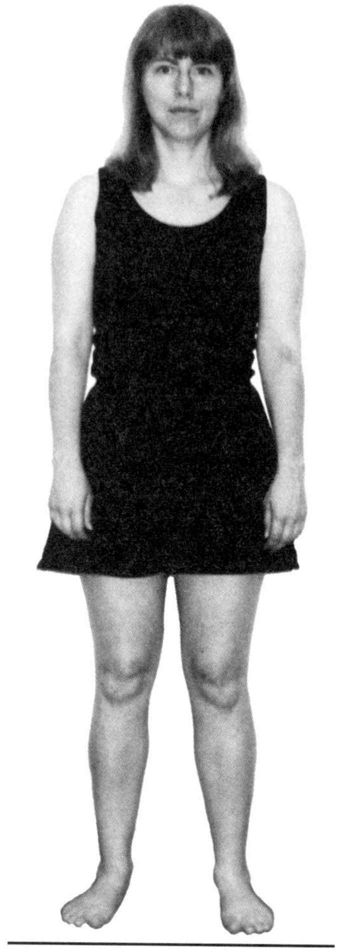

Notations of Weight Bearing on the Foot

Placement of all the body segments will determine whether one is centered more over the left or the right foot.

This can be notated by placing an arrow under the area that seems to be the center point of weight bearing.

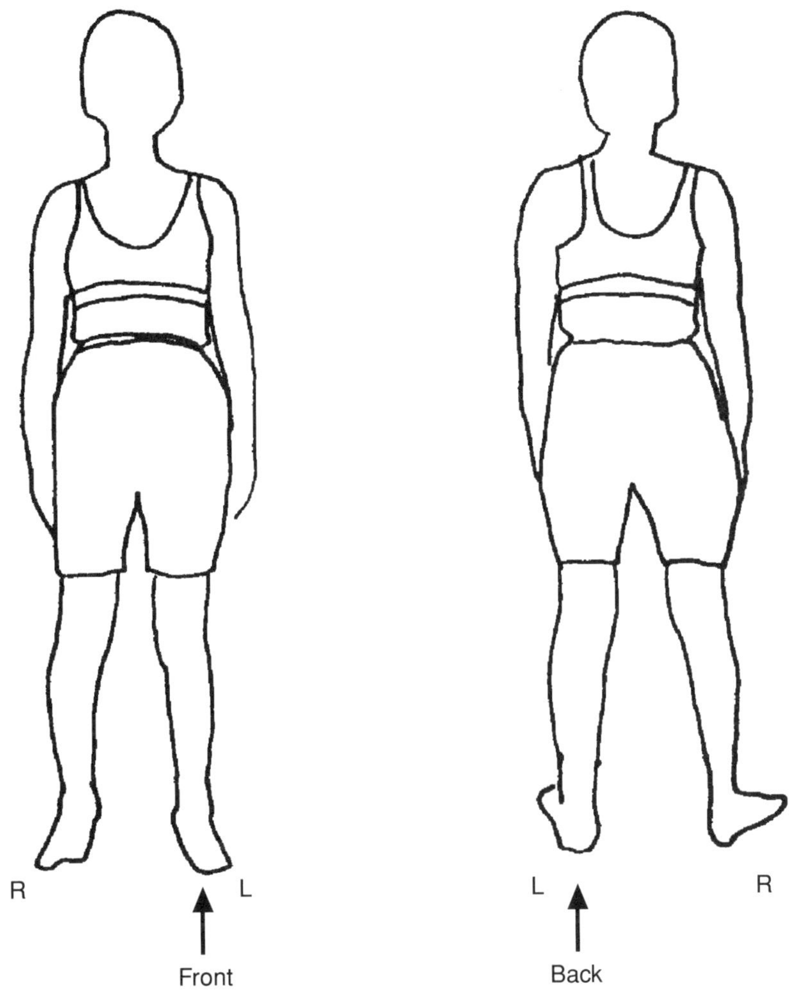

Exercise 2.25

Notate the appearance of the greatest weight-bearing area on the foot.

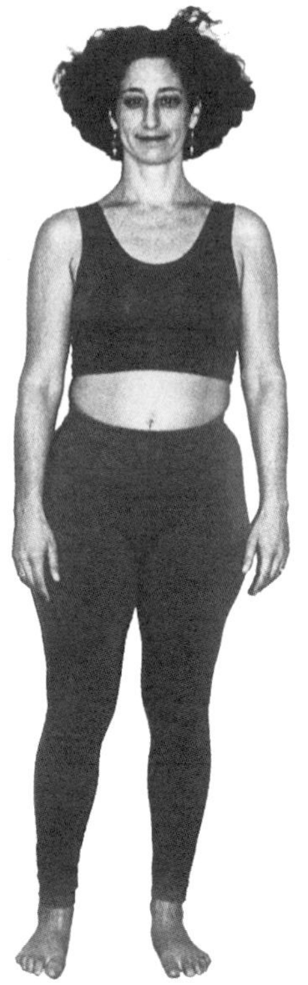

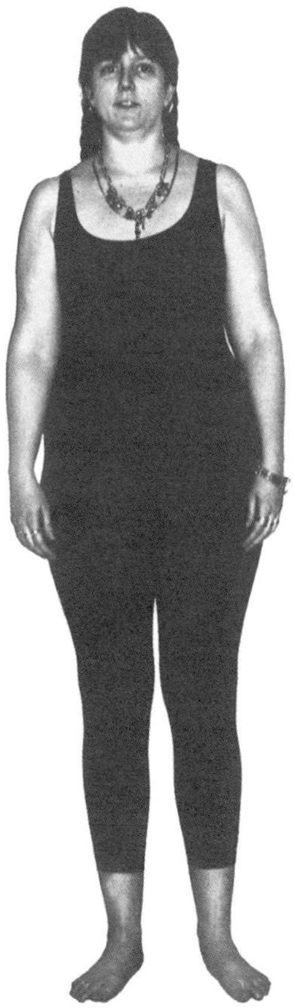

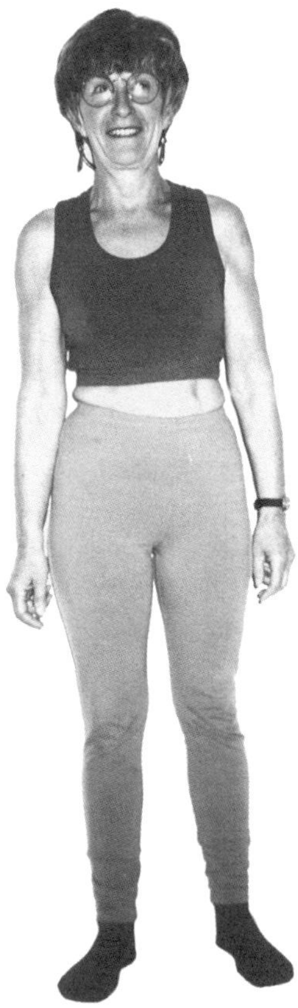

Take the following steps to help you determine the greater weight-bearing side.

1. Instruct your client to stand on both feet and shift his or her weight to the right side, and maintain this position until he or she feels centered and balanced. Repeat by having the client shift his or her weight to the left side.
2. Have the client shift his or her body weight to the right foot until he or she can pick up the left foot and then come back to the centered position.
3. From the center position, have the client shift his or her weight over to the left foot until he or she can pick up the right foot and then come back to the center.
4. Repeat this weight-shifting process until the side to which the client needs to move more is clearly established.

Generally, the client will move less to the side over which he or she is centered. There sometimes is a different pattern of weight bearing in the standing position than when lying down. For example, a person may have greater mass aligned over the left side while in the supine position and yet, in the standing position, have weight-bearing predominately on the right side. Sometimes this occurs because the person has a shorter right leg and in the standing position is being pulled to the right.

Use dental floss to see the plumb line and the right-to-left placement on the following three photos. Does the weight seem to be more centered to the left or the right?

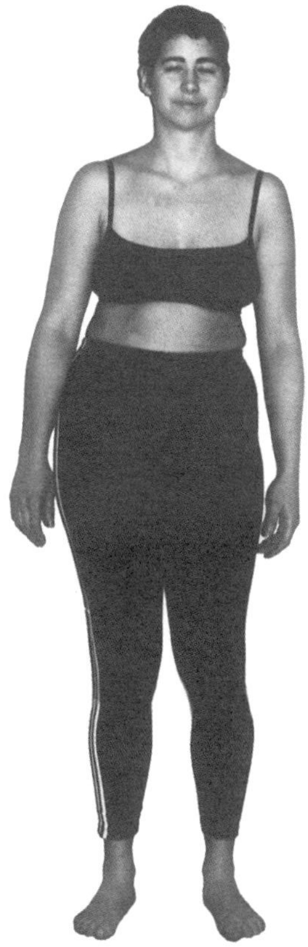

A

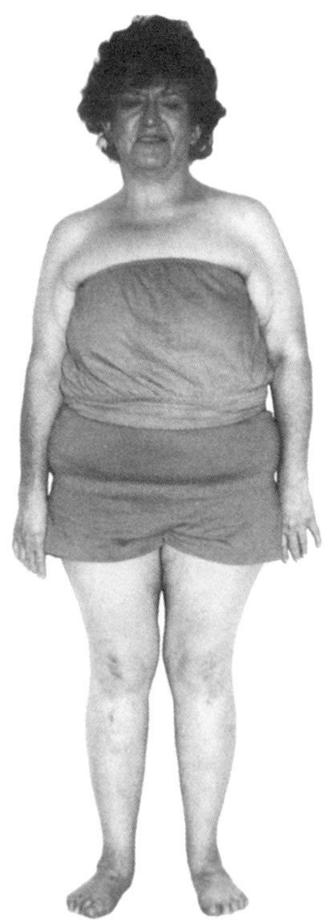

B

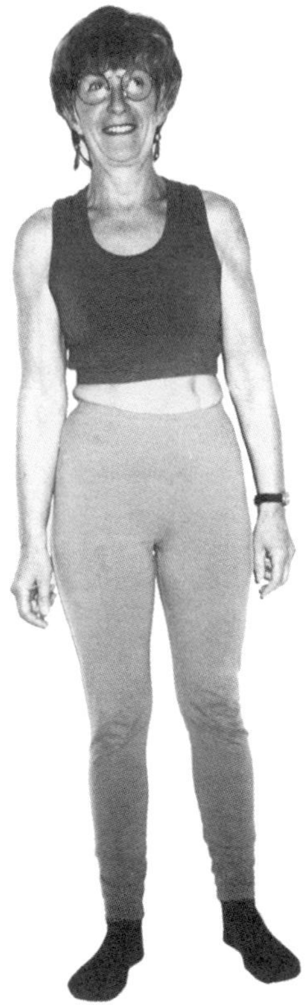

C

Symbols

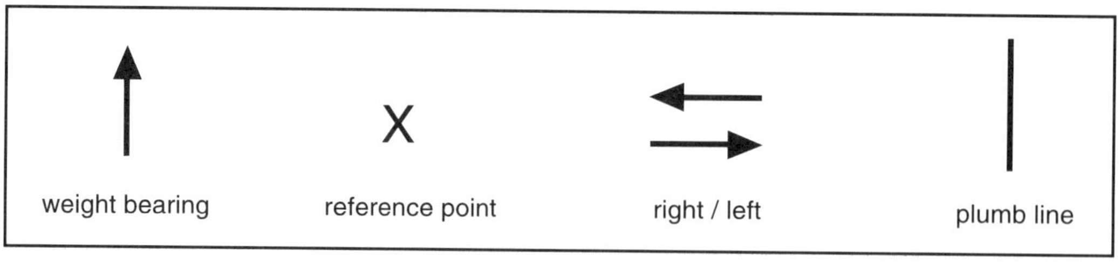

Exercise 2.26

Observe these three bodies. Outline the photos on a transparency.

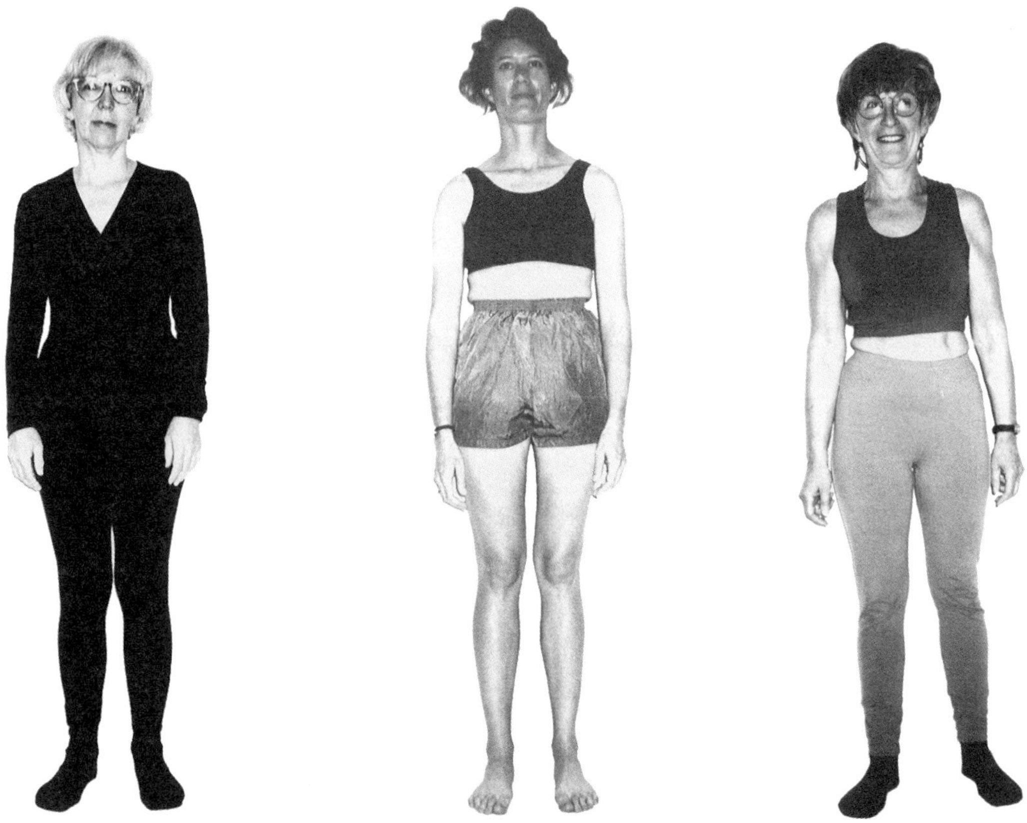

Exercise 2.27

Notate:

1. the greatest weight bearing. Is the up arrow closer to the right or left foot?
2. the plumb line
3. an X for the reference point
4. placement (right and left)

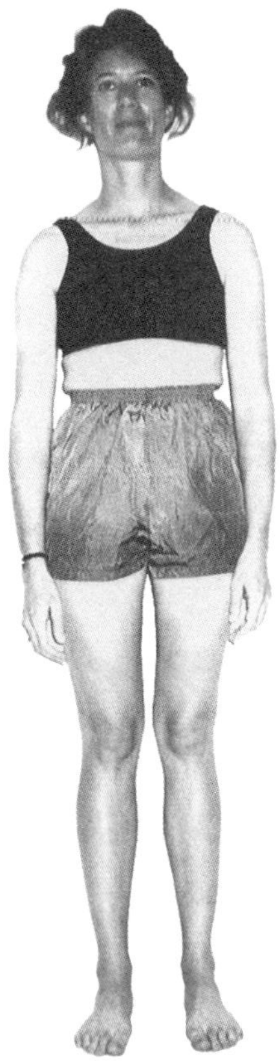

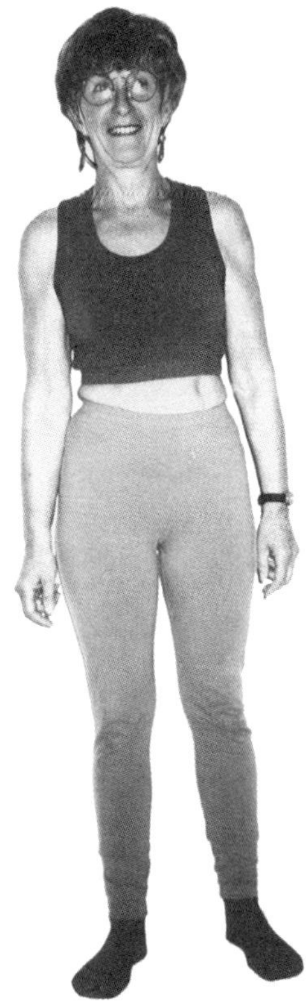

Now transfer your notations onto the body chart. Use arrows to designate position in relationship to the plumb line. You also may want to use the transparency to work with your own client photos. Place a transparency over the photo. Identify and notate the segmental relationships within the body. Remove the transparency and try to see these relationships without using the notation marks. This *seeing* ability is important to have when looking at an actual client in front of you. The notations are for your memory afterward and for studying a pattern later.

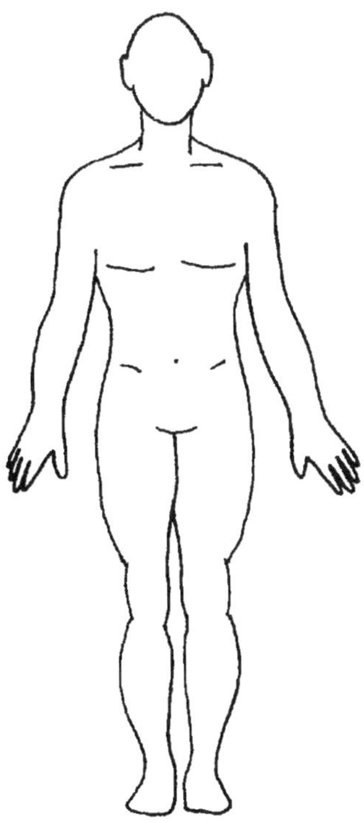
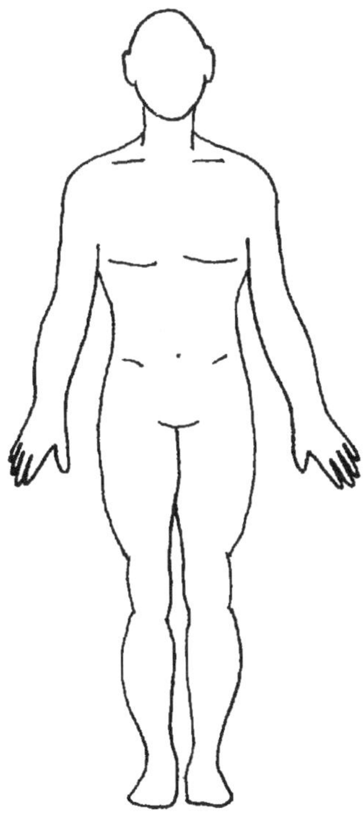
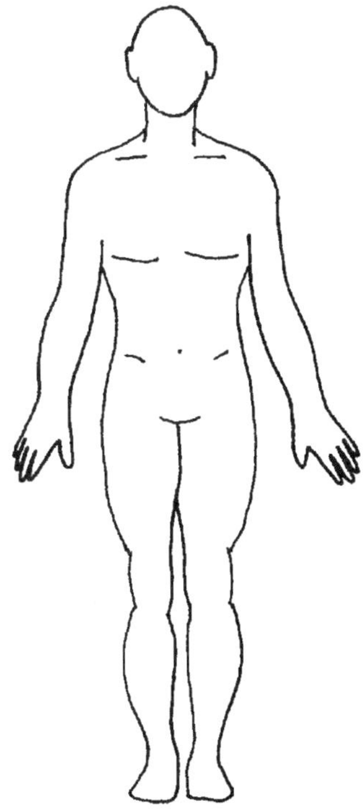

Rotations in the Frontal Plane

Exercise 2.28—Identifying Rotations

Mark the segmental centers directly on these photos and notice the relationships of the segments to the plumb line.

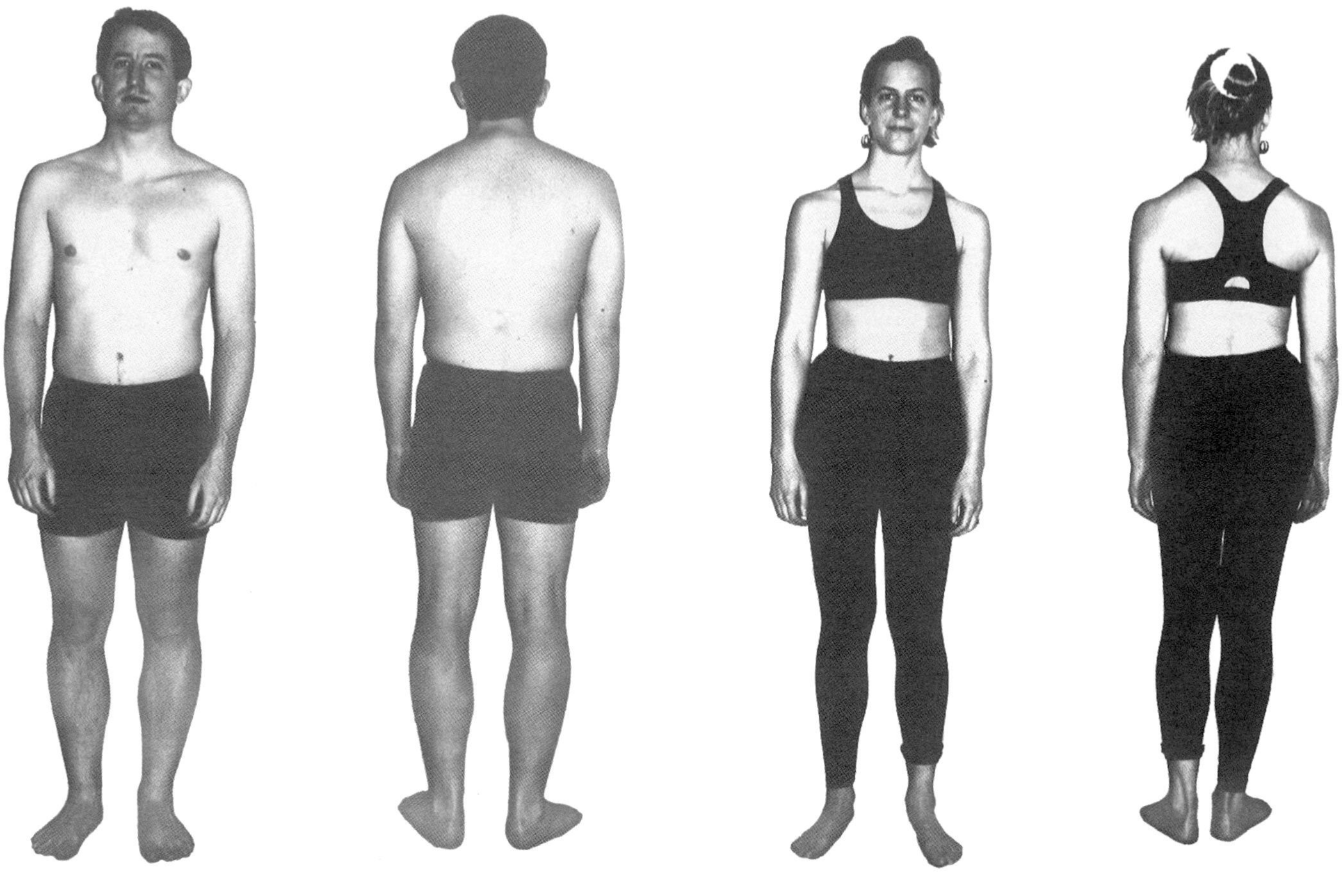

Do these segments appear shifted to the right or left only, or can you see the right or left side rotating forward or backward?

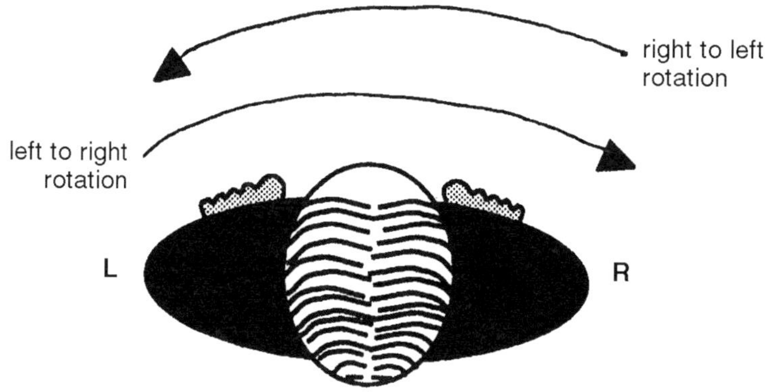

Notation of Rotation

The process of notating alignment in this plane is the same as in the front-to-back view. Arrows indicate displacement of individual segments to the right or the left. Curved arrows notate anterior/posterior placement, as well as internal/external rotations.

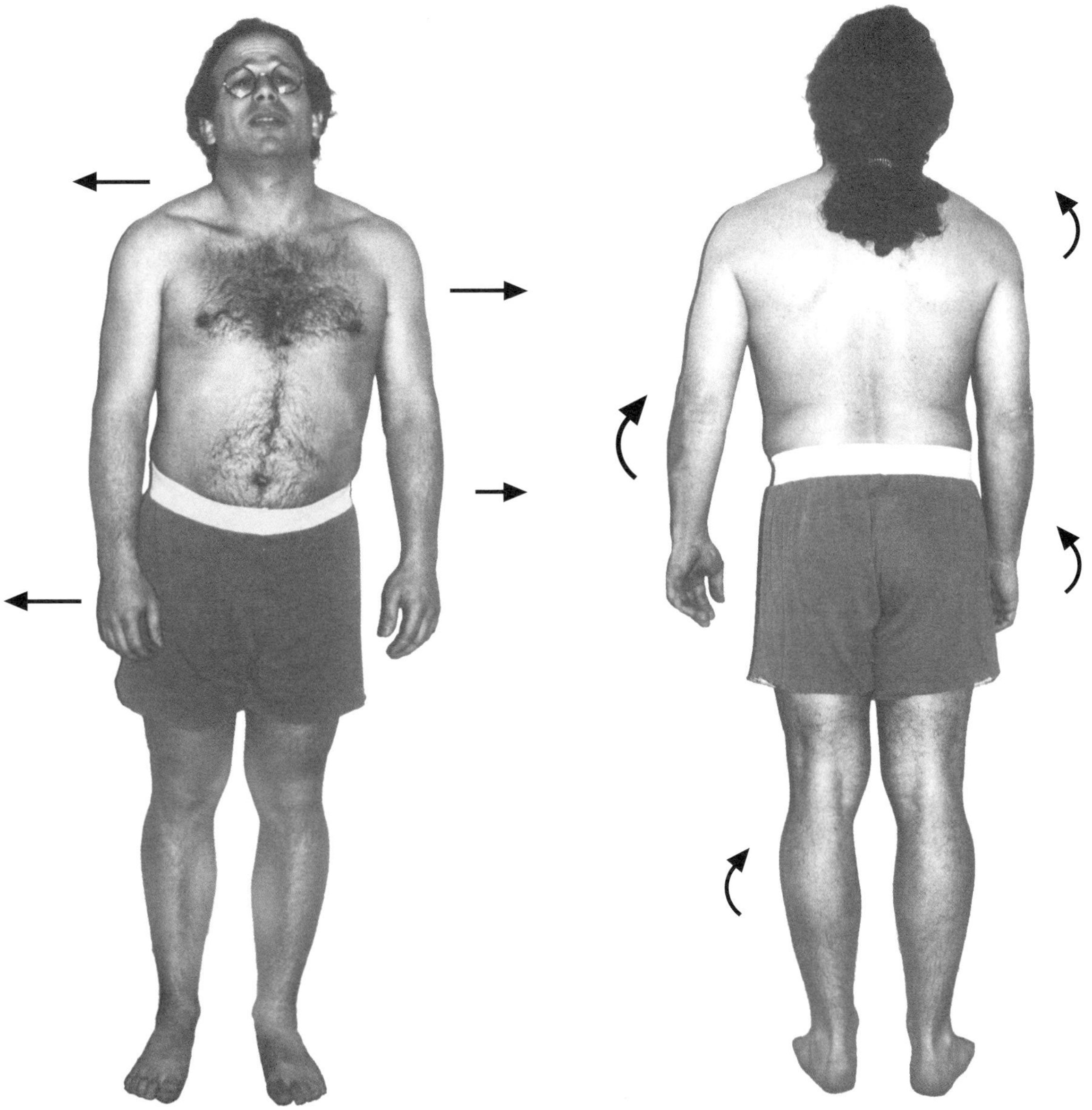

To indicate whether the segment is placed forward of (anterior) or behind (posterior) the segment above and below it, draw a vertical line beside the segment, as with this example, designating that the pelvic segment is forward on the left side (a left-to-right rotation).

1a. The symbol for one segment rotating forward

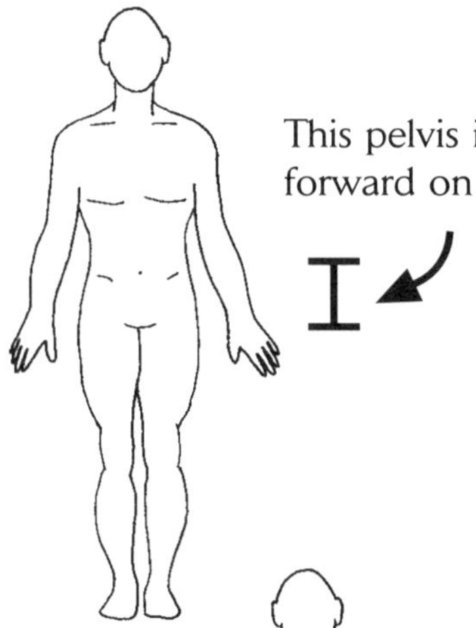

This pelvis is rotated forward on the left.

1b. The symbol for several segments that are rotated forward

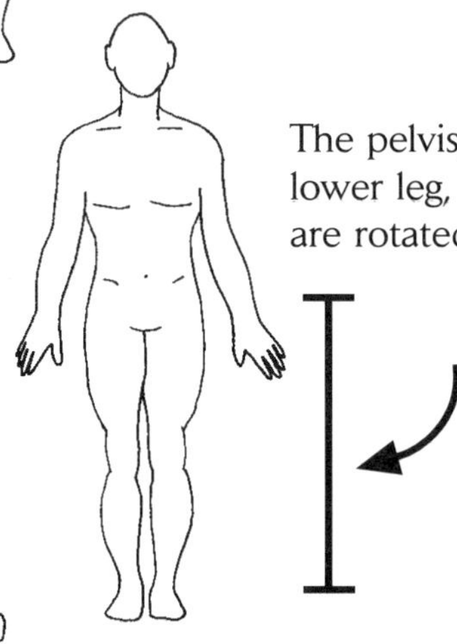

The pelvis, the upper and lower leg, and the foot are rotated forward.

If the joint is internally or externally rotated, place the curved arrow at the joint.

2a. The symbol for one segment rotating internally or externally

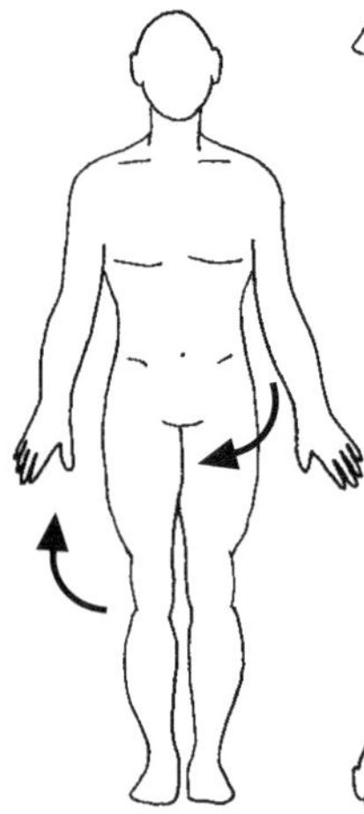

The left hip is internally rotated. The right knee is externally rotated.

2b. The symbol for several segments internally or externally rotated

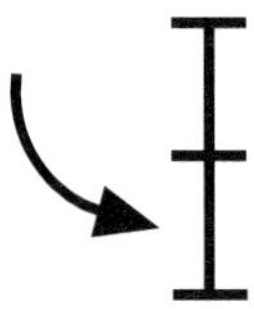

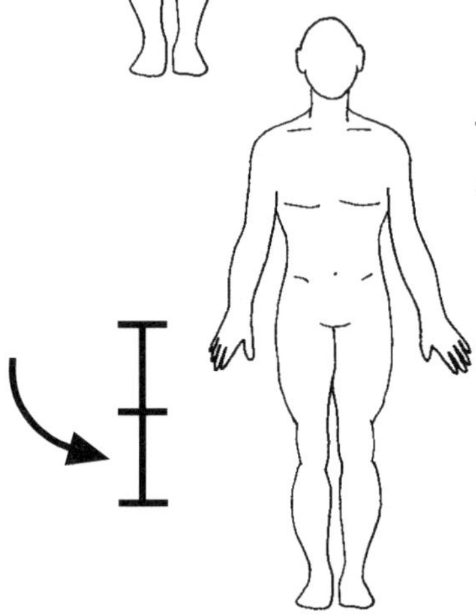

The right hip, knee, and ankle are internally rotated.

Demo

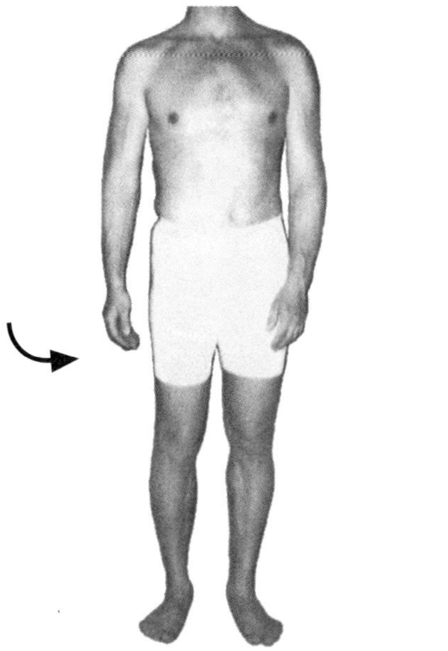

Right-to-left pelvic and
upper leg rotation

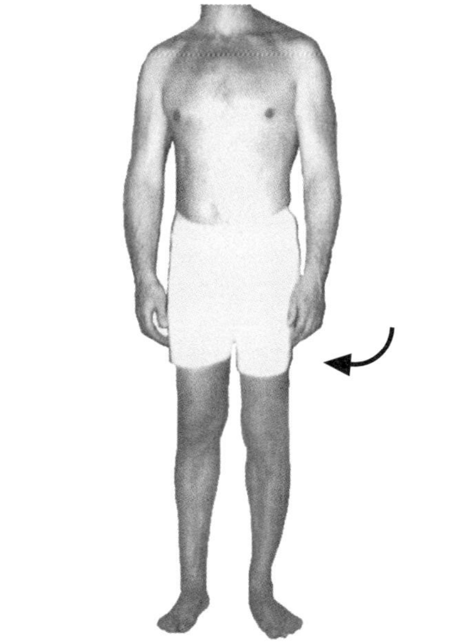

Left-to-right pelvic and
upper leg rotation

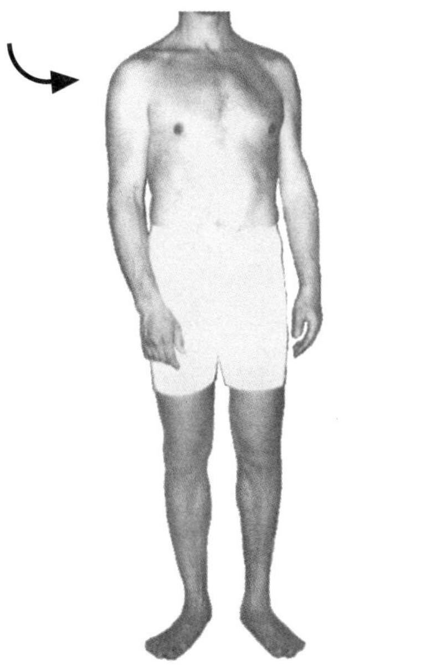

Right-to-left rotation of
mid- and upper trunk

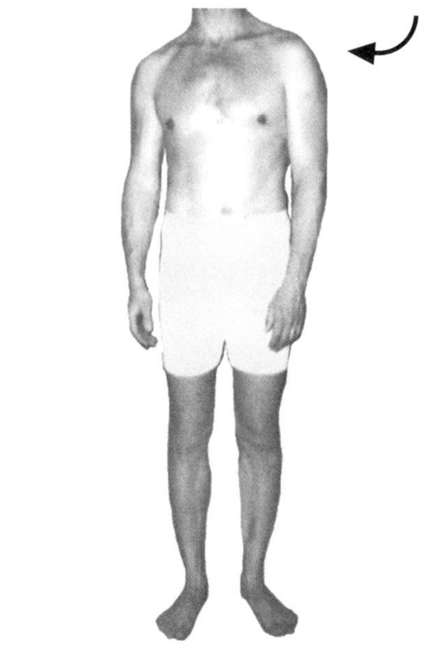

Left-to-right rotation of
mid- and upper trunk

Exercise 2.29

Notate the rotations indicated in the following bodies.

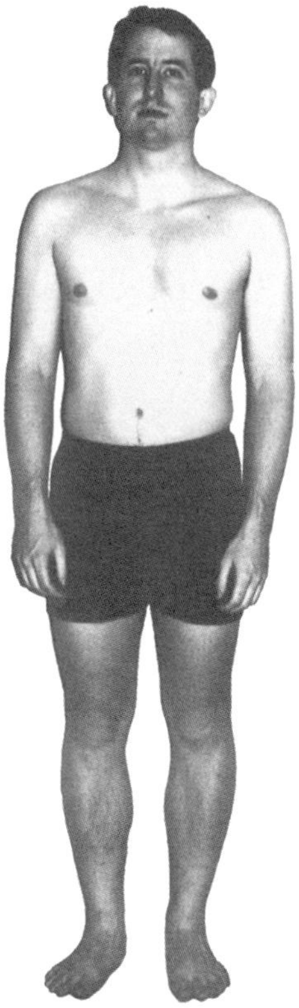

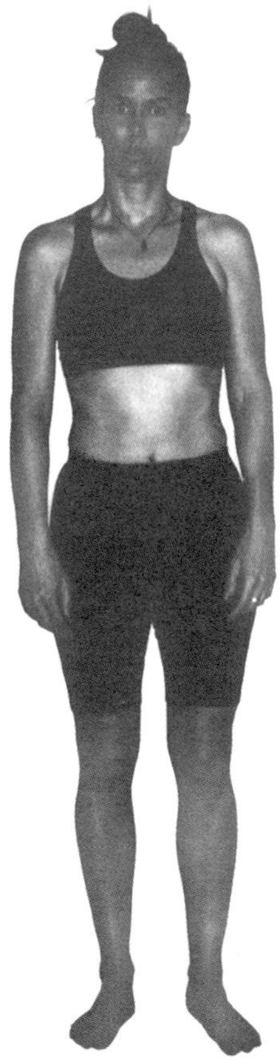

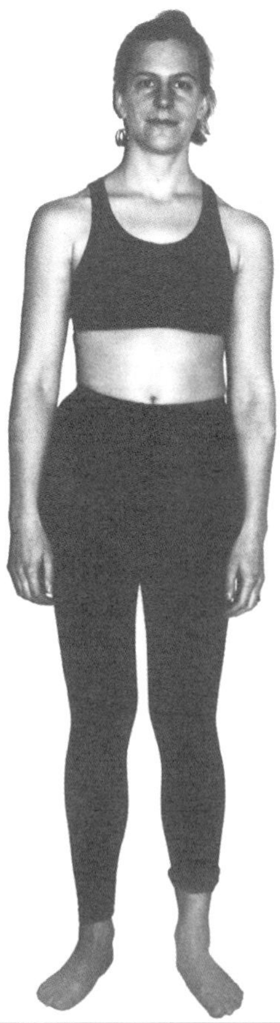

Seeing Alignment in the Transverse Plane

Landmarks—Front View

Observing the frontal plane, we just looked at right and left displacements and rotations. Now, in the transverse plane, let's identify the high-to-low (up to down) position of the different landmark areas.

Landmark Options

1. top of head or ear
2. temporomandibular joint (TMJ) or ear*
3. jaw (inferior border)
4. top of clavicle or a cross sternal notch**
5. shoulder joint
6. costal arch
7. iliac crest
8. hip joint
9. knee joint
10. ankle joint
11. elbow joint
12. wrist joint
13. tips of fingers

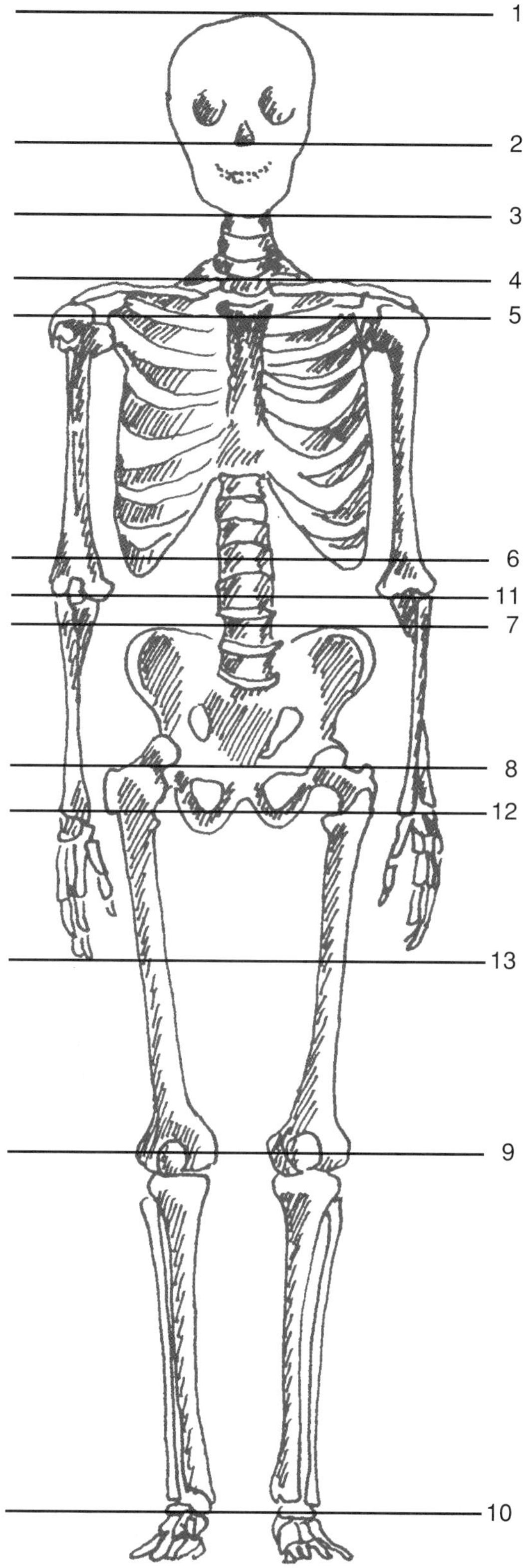

*Most times you see the angle at the ear and the temporomandibular joint (TMJ) to be the same; however, with detailed observation, you may identify the ear as level, but unevenness at the TMJ reveals a strong asymmetry within the joint.

**Sometimes you can observe the clavicle low on one side and the humerus high on that same side.

Connecting Landmarks to Observe the Up-to-Down Relationships

Exercise 2.30

Using a transparency:

1. On the second skeleton, below in the middle, mark the landmarks.
2. Draw the lines to connect the landmarks.
3. Observe the line drawing on the far right, which is the same person as the second skeleton.

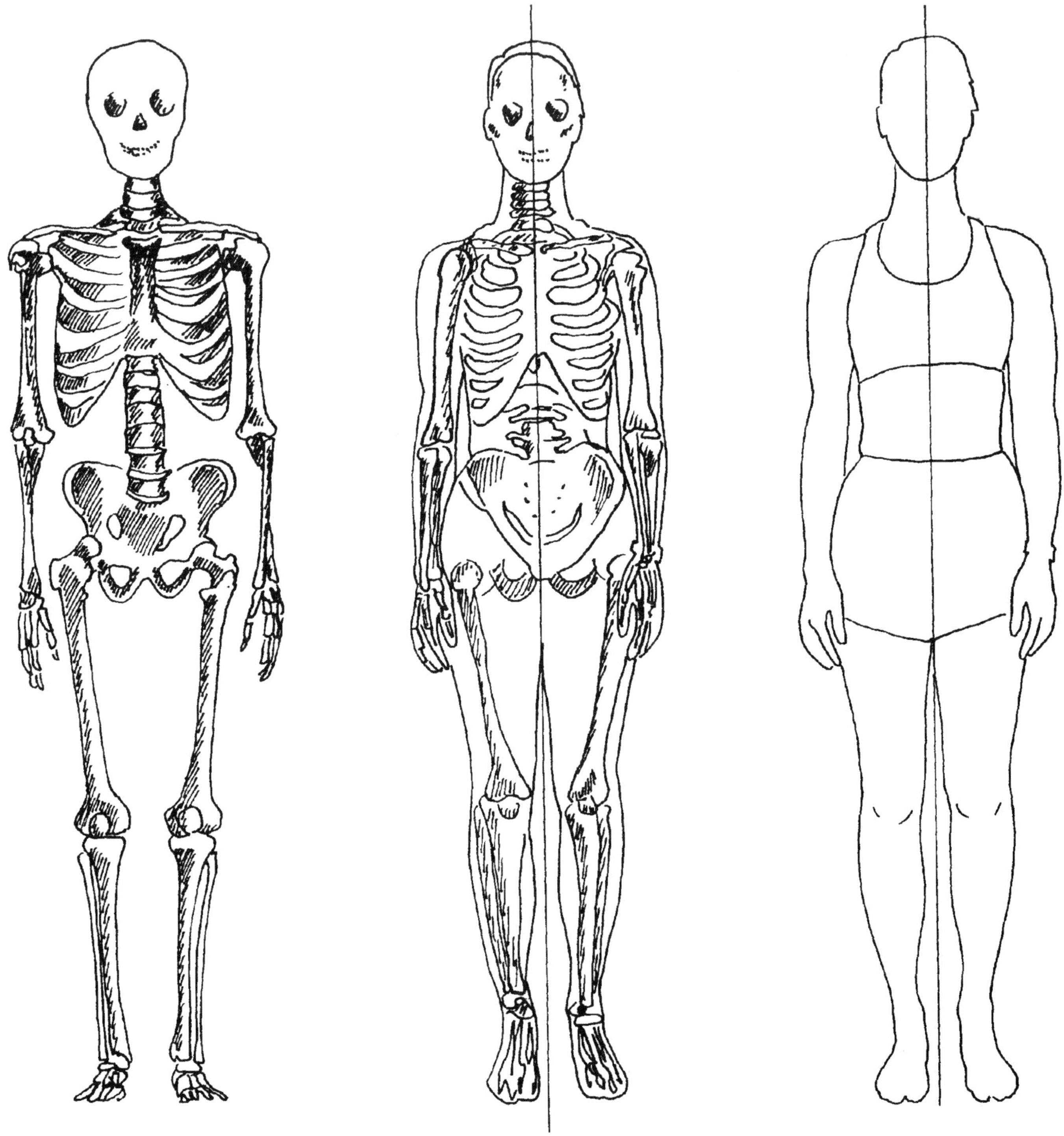

Now, let's observe a body pattern that is not as symmetrical as that in the last illustration. We have notated the landmarks demonstrating more obvious displacement.

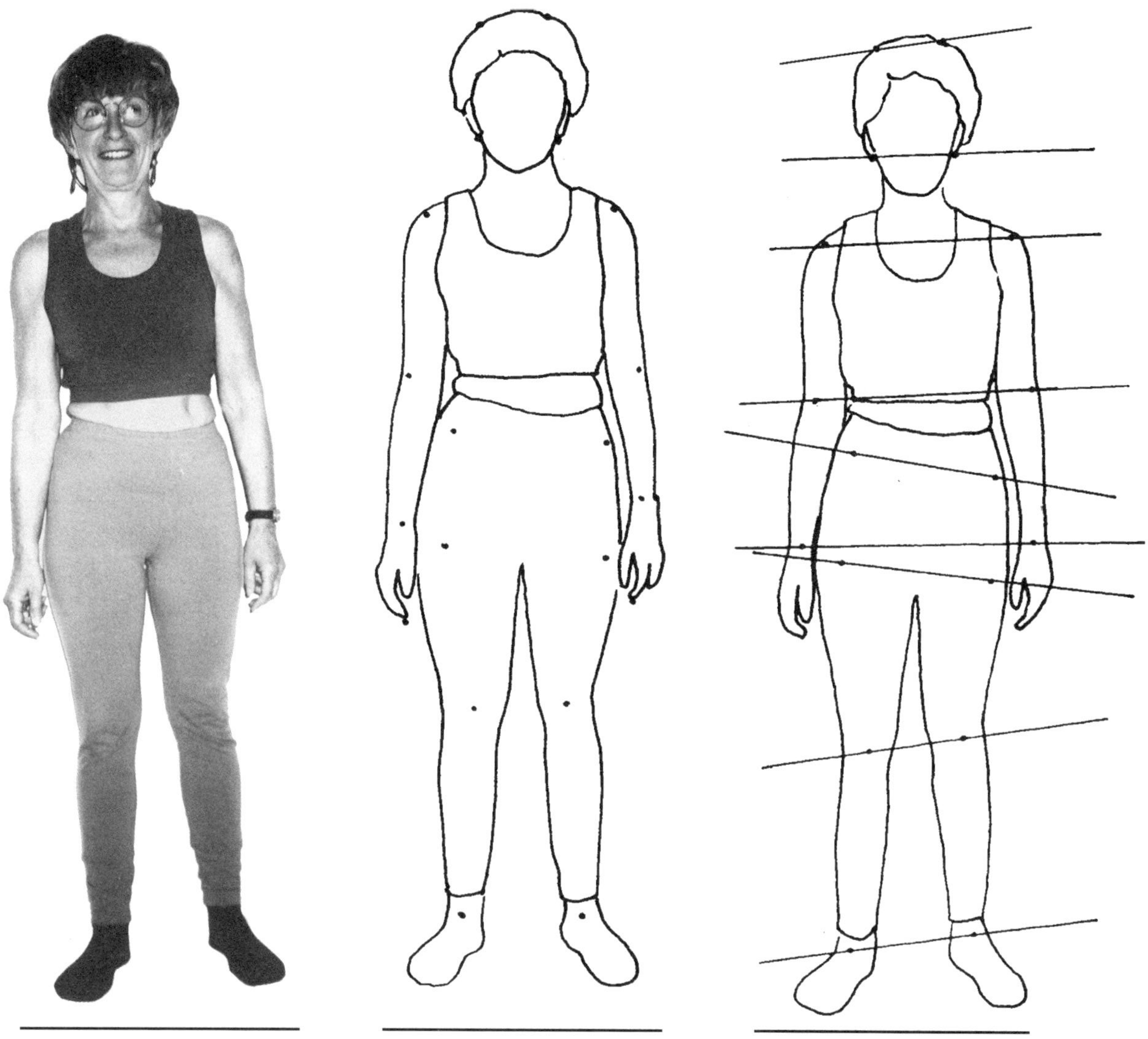

Exercise 2.31

Draw landmarks and lines connecting the reference points. Observe the high-to-low relationship.

Exercise 2.32

Draw the landmarks and the lines that connect them.

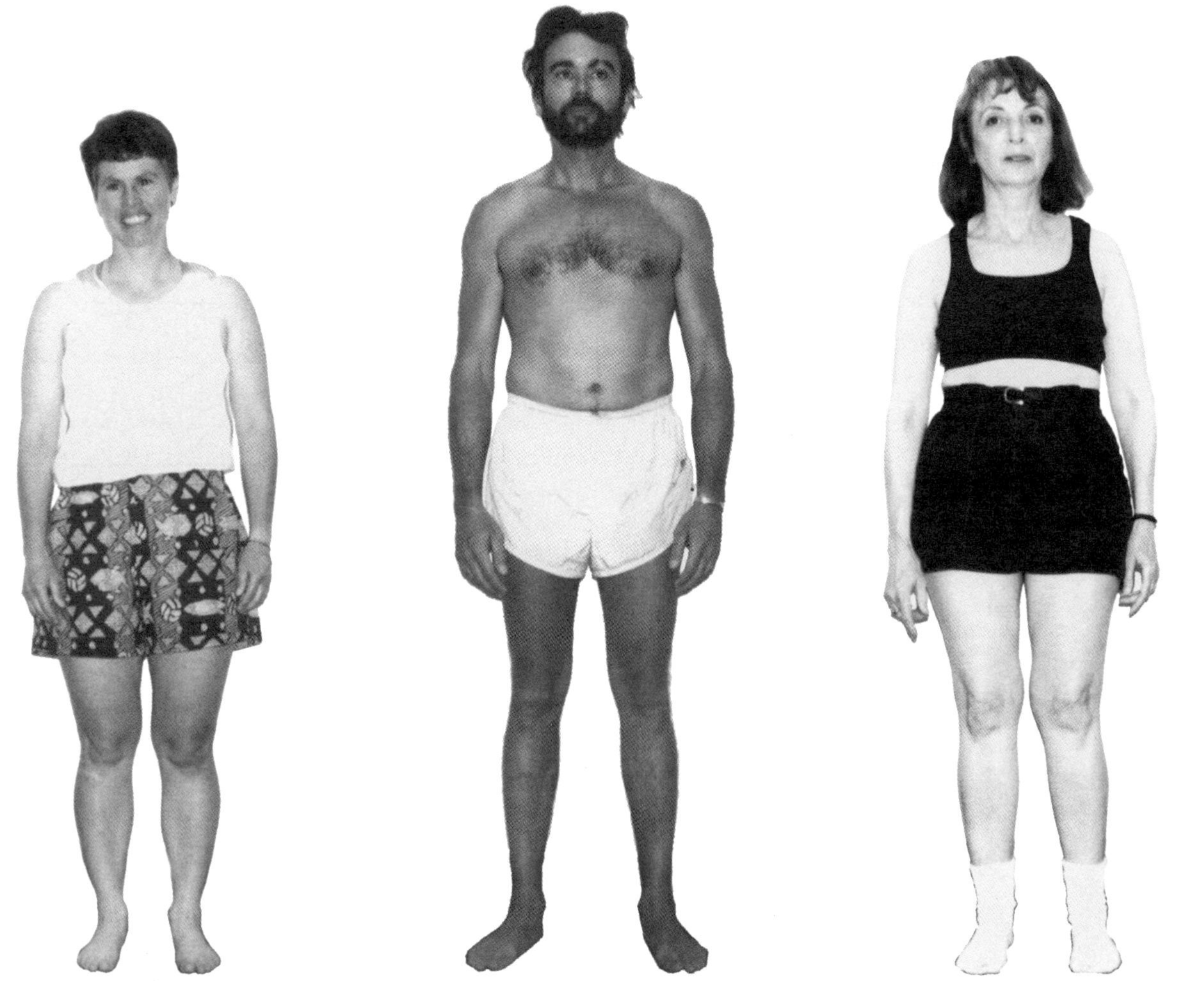

Landmarks—Back View

1. top of head
2. ear
3. occipital ridge
4. shoulder joint
5. inferior tip of scapula
6. elbow joint
7. iliac crest
8. hip joint
9. wrist joint
10. tip of finger
11. knee joint
12. ankle joint

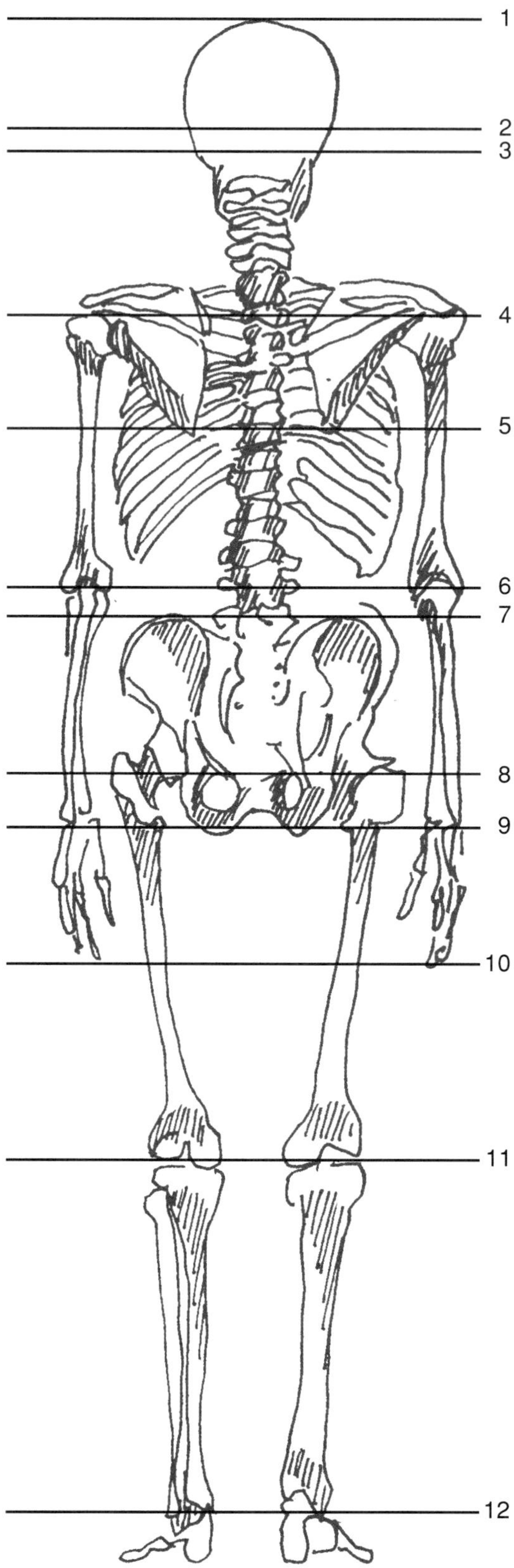

Connecting Landmarks—Back View

On a transparency, mark the landmarks across these figures.

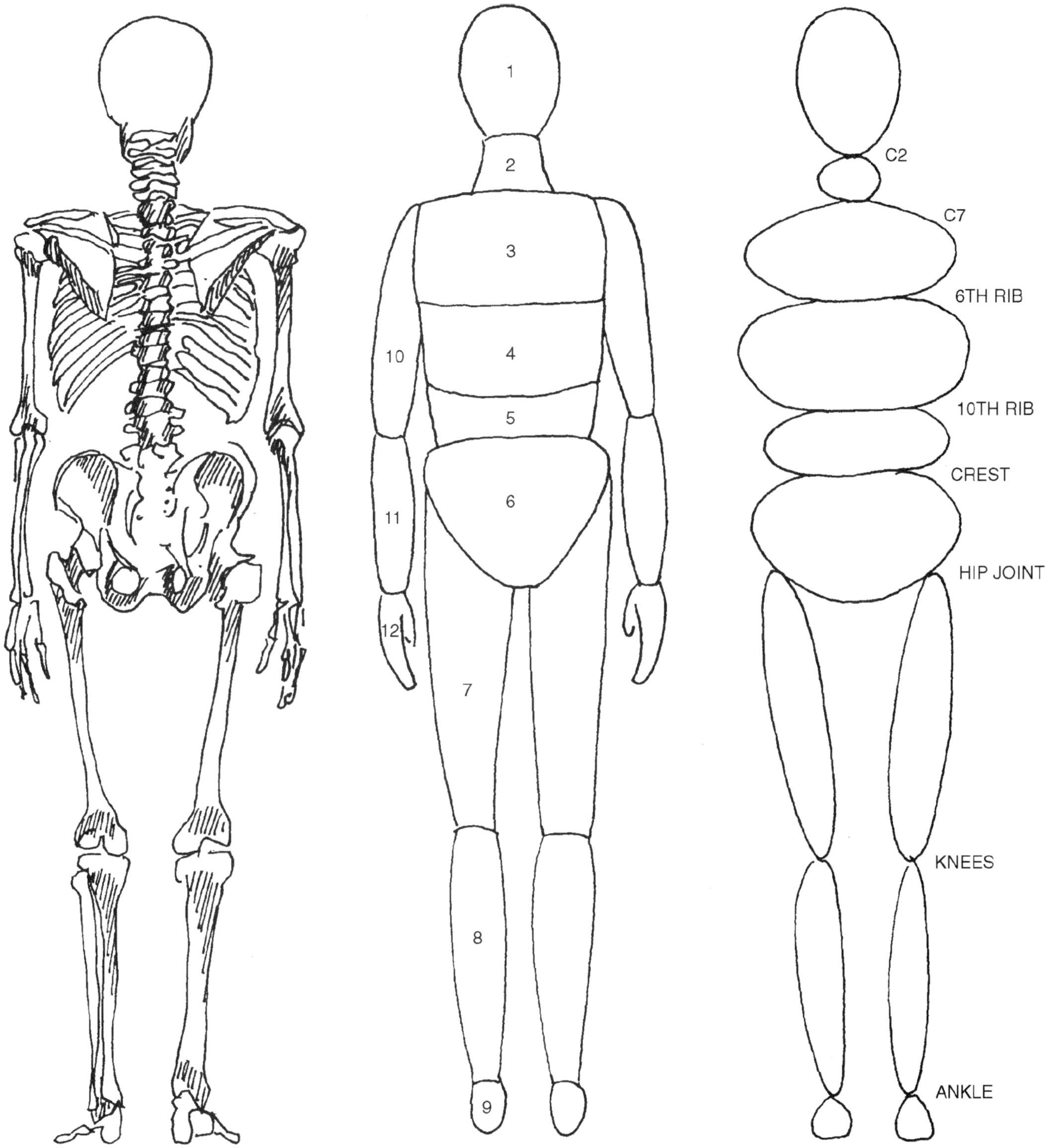

Demo Figures

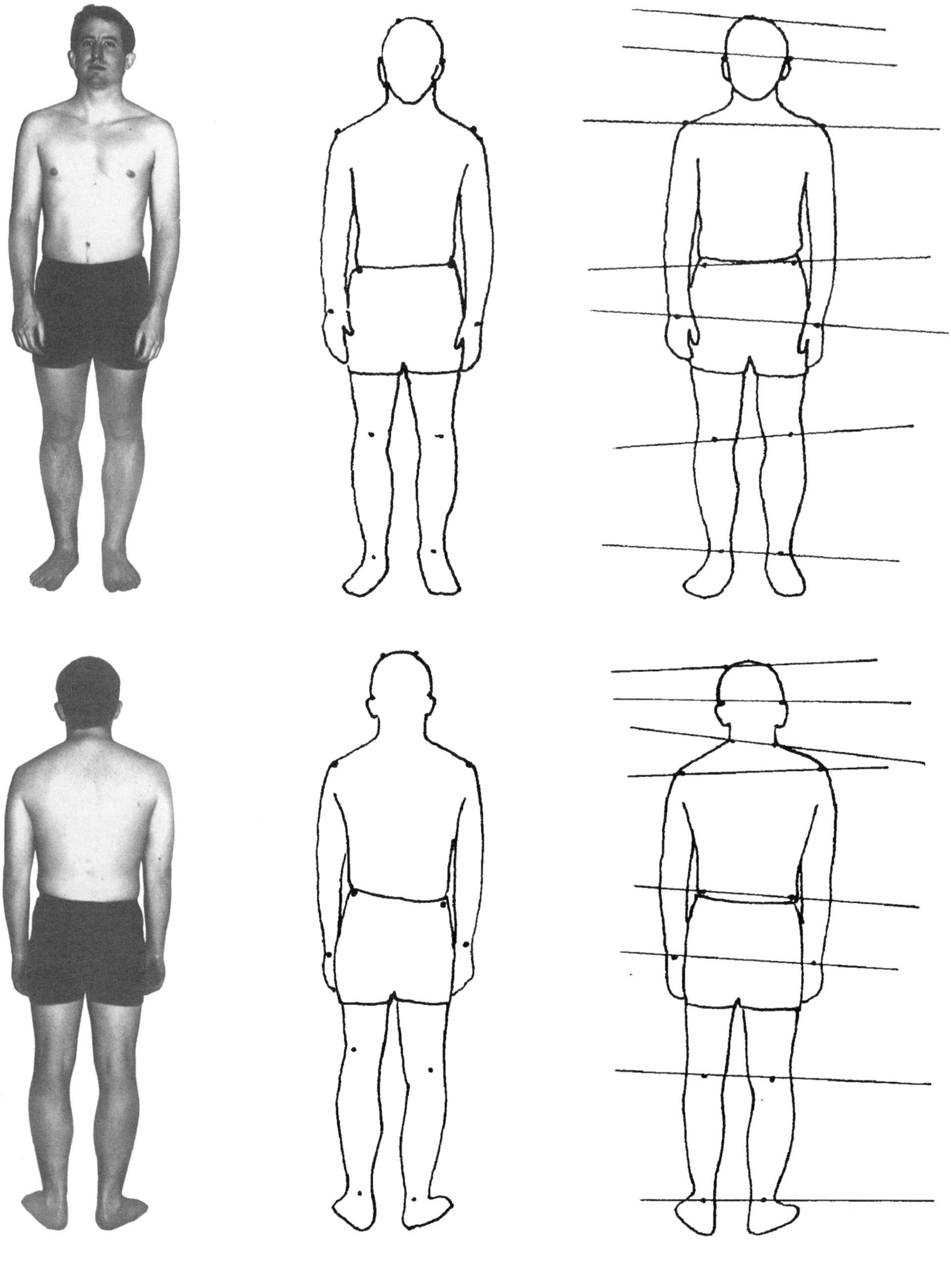

Observe the front and back views. Are the high-to-low (up-to-down) patterns obvious from both sides? Are they different from one view to the other?

Exercise 2.33

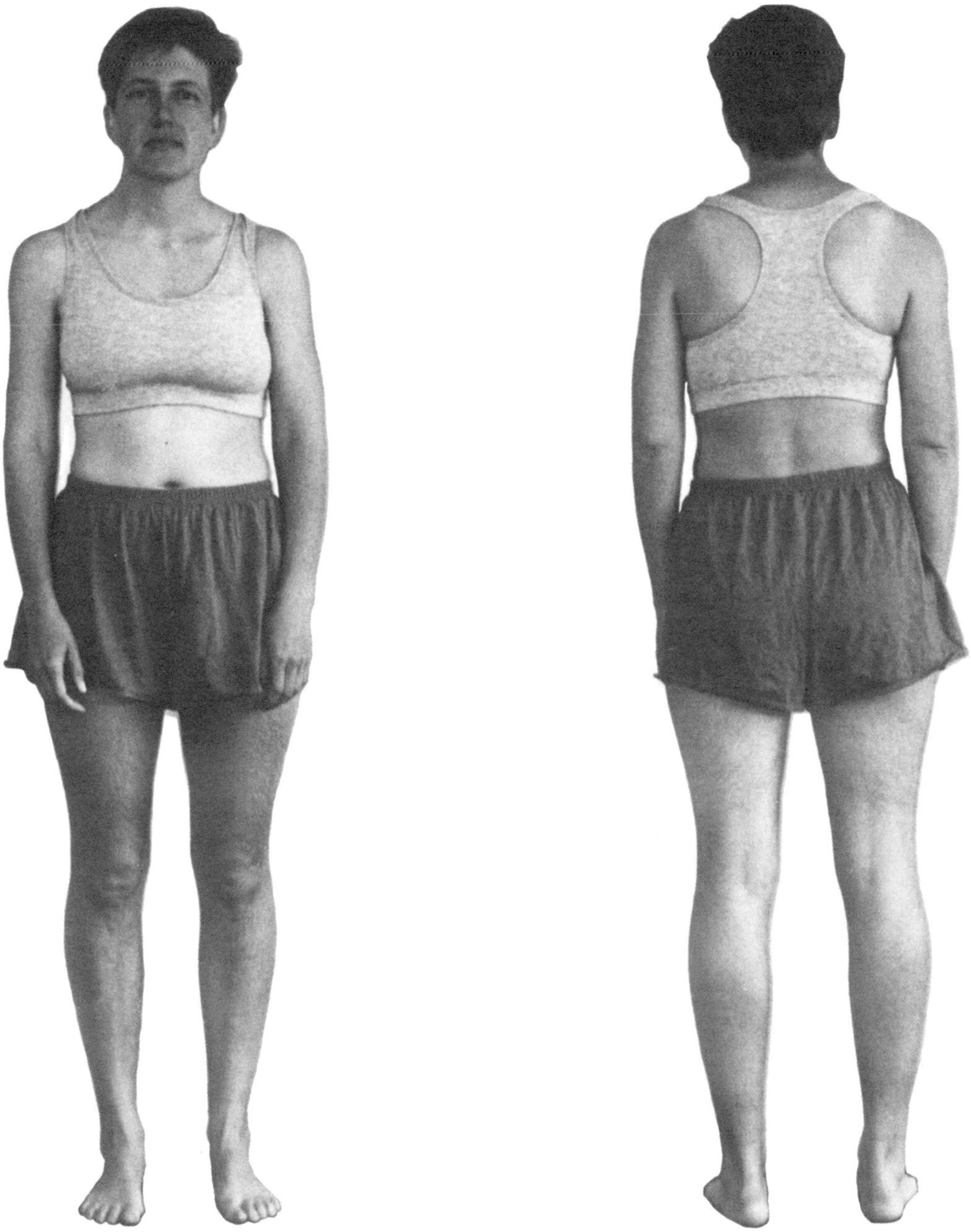

On a transparency, mark and connect the landmarks. Now, place your notations beside the photos and compare your ability to see these relationships in the photo without your transparency notations.

Exercise 2.34

Notate the segments that most get your attention.

1. up-to-down placement
2. right-to-left placement
3. rotation
4. the greatest weight-bearing area on the right or left foot

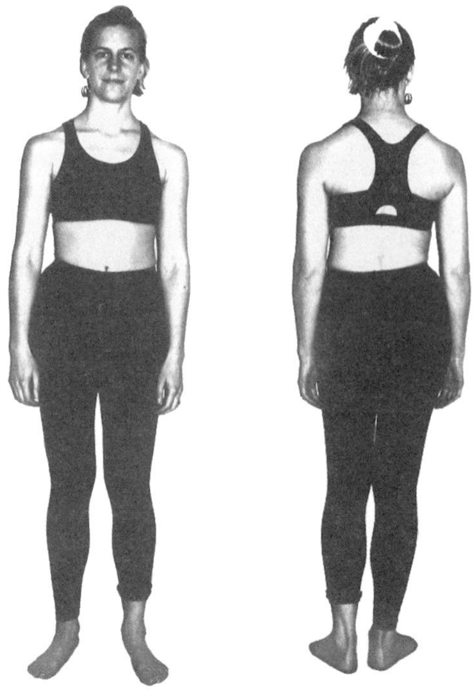

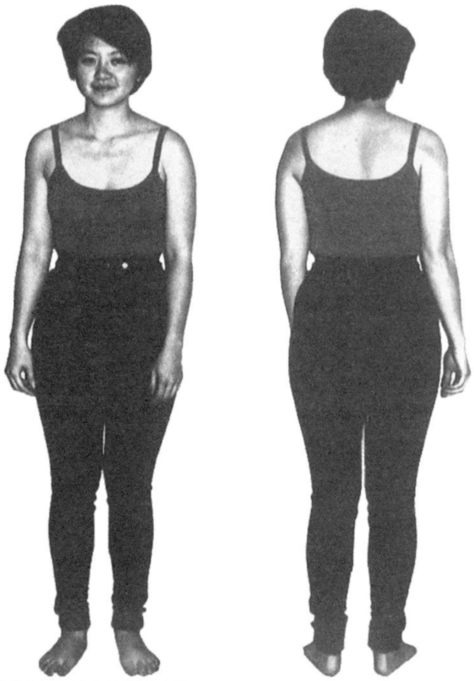

Placement of the Pelvis

We have just observed the up-to-down relationships of segments from the front and back views. Let us now examine the up-to-down relationship of the pelvis (the pelvic tilt) from the side views.

Earlier in this chapter, we observed the pelvis from the sagittal (side) view, with its anterior-to-posterior placement, and from the frontal plane (front and back), with the right-to-left placement and the high-to-low placement.

The difference between placement and tilt is an important and often overlooked distinction. Before we discuss some of the complex combinations possible, we need to review the characteristics of the center of mass (or center of gravity), and discuss a method for determining tilt.

First, it is important to note that the center of gravity of the pelvis is in the same place, regardless of the angle or tilt of the pelvis. One way of determining the center of gravity of an object is to suspend it from different points (which produces different tilts) and find the point where the gravity lines intersect.

In order to determine the tilt of the pelvis, we need to move the landmarks from the center of gravity to points at the top of the pelvis. We will consider the posterior/superior iliac spine (PSIS) to the anterior/superior iliac spine (ASIS) as the top landmarks.

The following examples show a more neutral pelvis as well as ones with an anterior and posterior tilt, respectively.

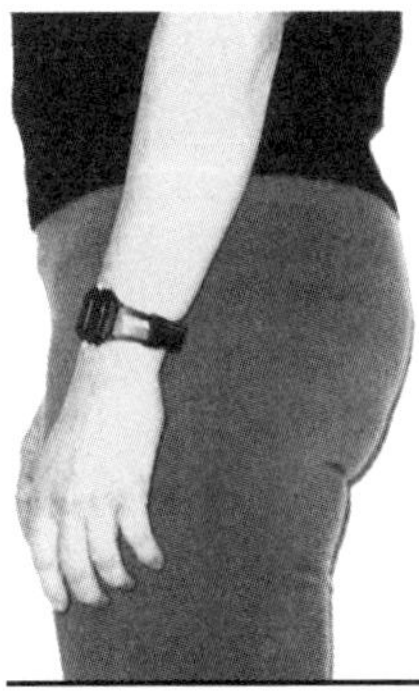

neutral

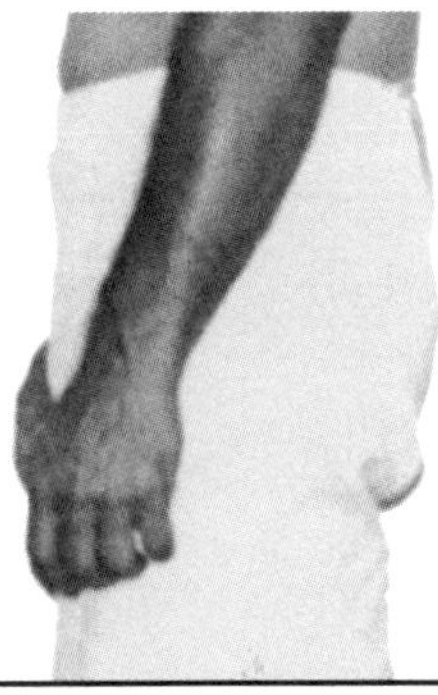

posterior tilt

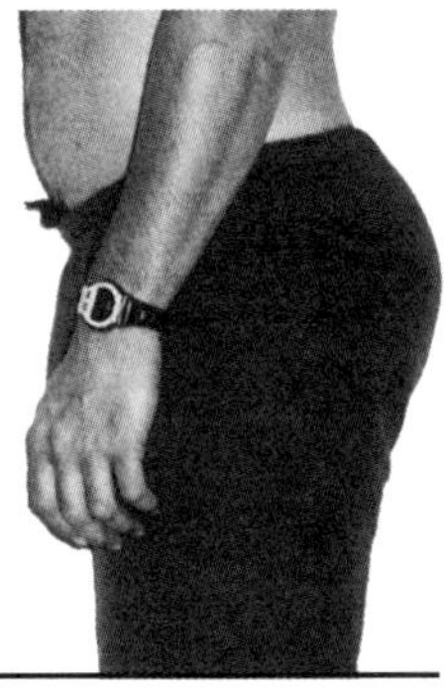

anterior tilt

Notation of Pelvic Tilt

Notate the tilt of the pelvis in these photos by drawing a line from the ASIS to the PSIS, and from the pubic symphysis to the sacral/coccyx junction.

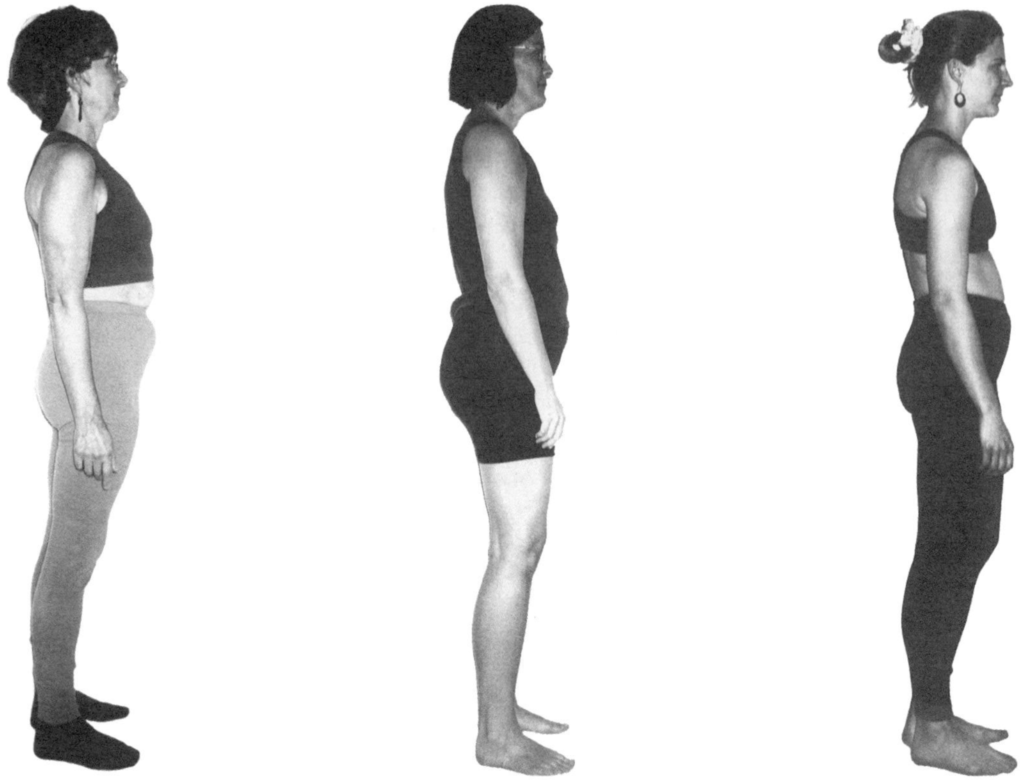

We now can transfer these diagonal lines to the abstract body chart to notate the angle of the pelvic tilt.

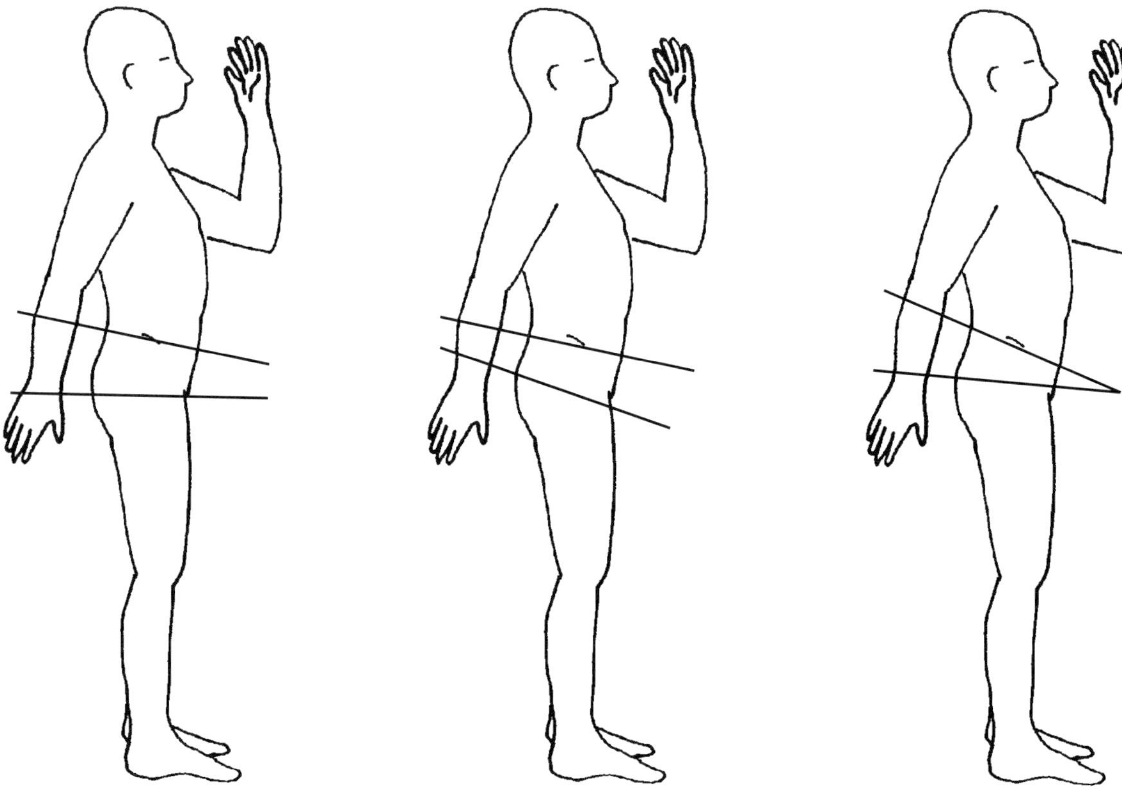

Tilt and Placement of the Pelvis/Hip

The placement of the center of gravity of the pelvis can combine with different tilts in many possible configurations. For example, each of the following figures shows that the placement of the pelvis is anterior to the plumb line. As you can see, however, the tilt of the pelvis is different in each case.

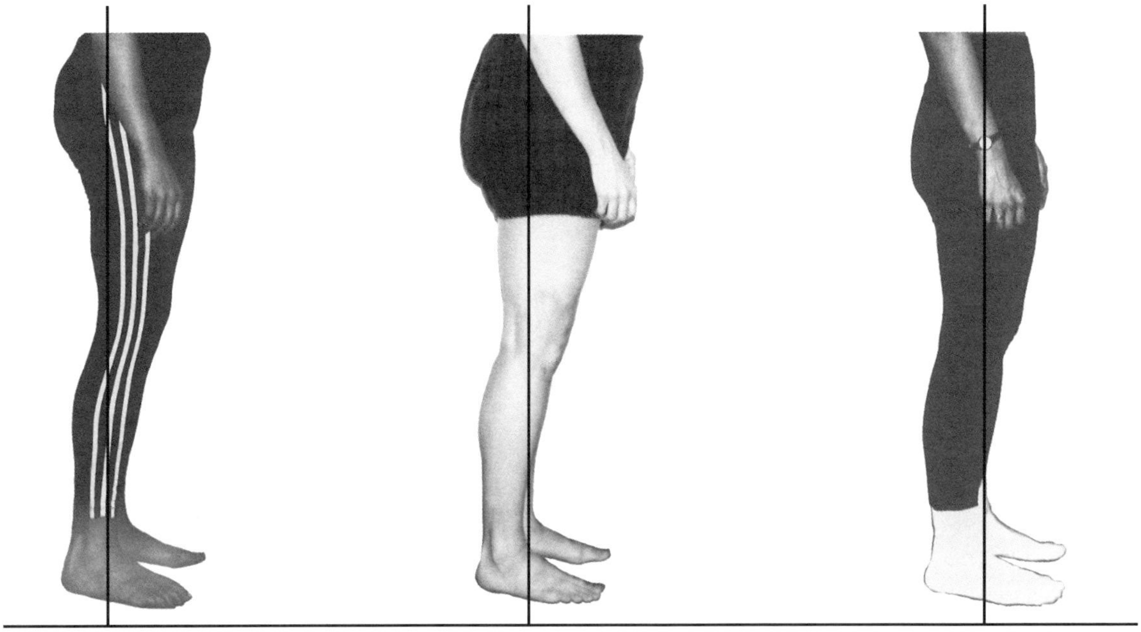

more neutral   anterior tilt   posterior tilt

Study the following examples, which indicate other possible combinations. Also, note the distance of the ischial tuberosity to the femur.

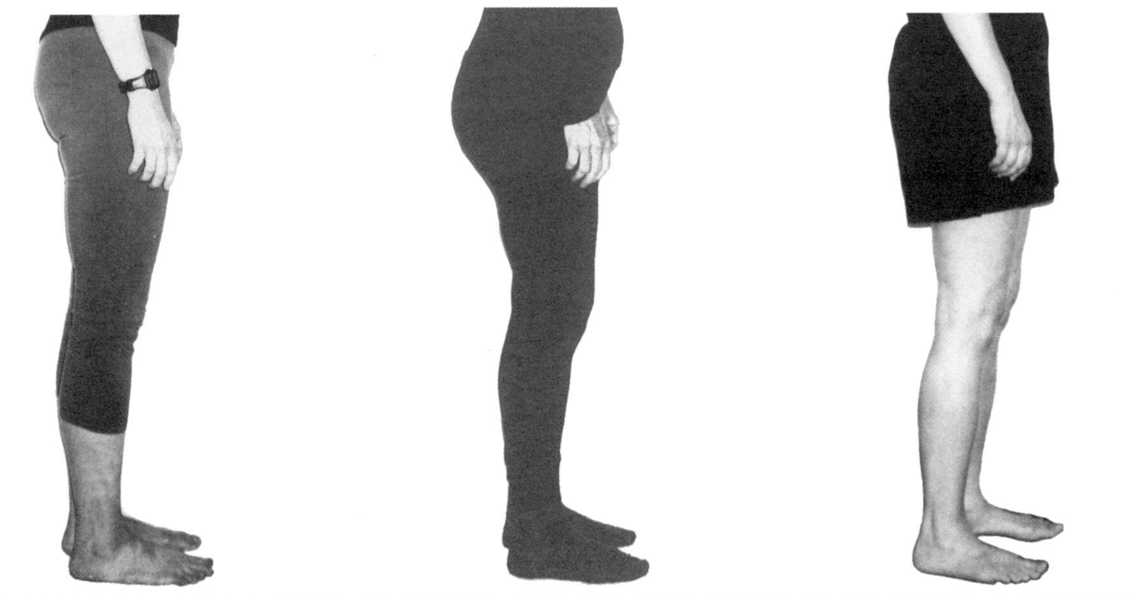

more neutral   anterior tilt   posterior tilt

Varying tilts and placement

Varying Tilts and Placement

On a transparency, notate placement of the pelvis and hip, as well as pelvic tilt.

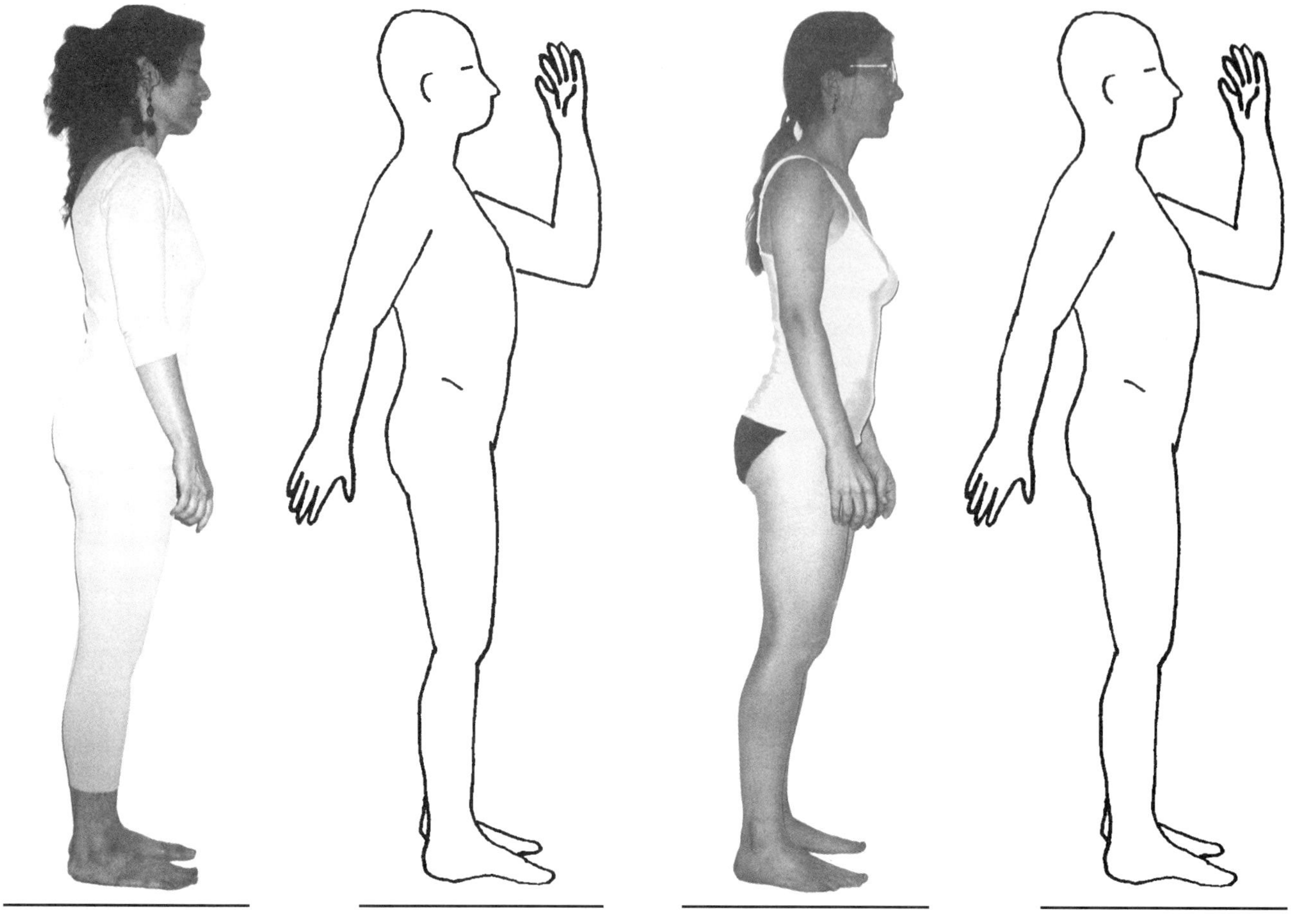

Remember that the X notates your reference point. An X at the ankle notates whether the hip joint is anterior or posterior to the ankle. Mark the combined placement and tilt on the abstract body chart.

Implications

As you can see, the possibilities for the combinations of placement and tilt are many. This becomes important in determining how to suggest a modification to the client. For example, a common correction given to a person with an anteriorly tilted pelvis is to "tuck the pelvis under." If a client has an anteriorly tilted pelvis that is anteriorly placed, he or she may not be able to do this successfully without first bringing his or her pelvis to a more neutral alignment. You may want to try this example with your own body. Conversely, a person with a posteriorly placed pelvis, a posterior tilt, and a decreased lumbar curve may not be able to normalize that curve without adjusting both the tilt and the placement of the pelvis.

Pelvic Tilt on Right and Left Sides

Observe your previous outline of the body segments on this photo.

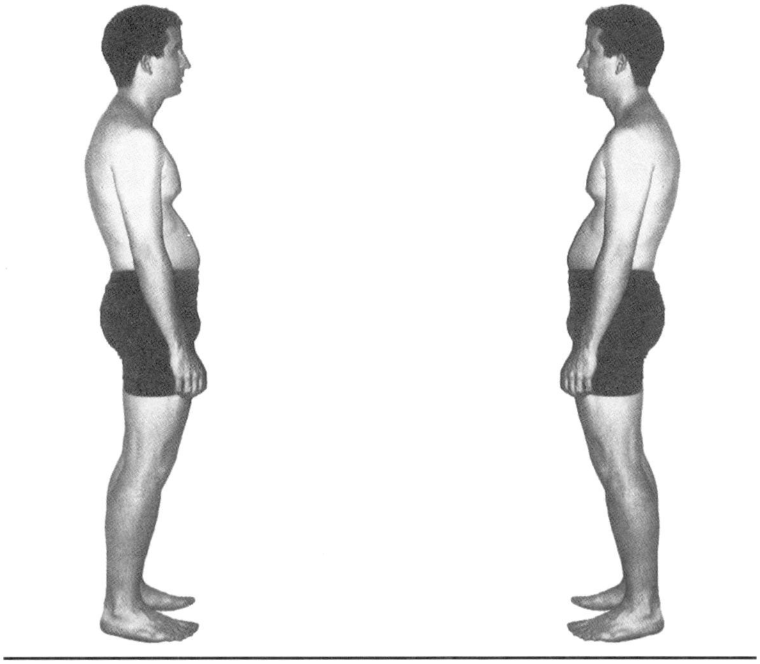

His right side **His left side**

Compare his two sides. They do not appear the same.

We most likely are seeing the effect of rotation. We also might be seeing the possibility that the left ilium, which appears to be in a posterior tilt, could be placed lower (more inferior to the right ilium). The right ilium, which appears to be in a more neutral tilt, might actually be higher than (or superior to) the left ilium. In this case, the left ilium (down-slip) position in relation to the right ilium (up-slip) creates increased stress on the sacral area, which is forced to negotiate these different positions. Often, people do not think that a fall, such as one on the ice, could have much consequence. In some instances, the moment of just such an impact can force one ilium high or low.

Integrating the Three Planes for Observing, Selecting, and Prioritizing Information

The previous pages contain exercises detailing specific notations for your learning, but you can appreciate that putting all the symbols on a chart would be prohibitive time-wise and overwhelming in terms of the amount of information.

Prioritizing Information

You should observe the body from different views, noting which view gives you the most information. This generally is the view (plane) that reveals the most extreme displacement.

Use the following list to draw your conclusions from the client photos.

1. Identify the view that draws your attention.
2. Which segment within that view gets your attention the most?
3. What is the relationship of that segment to the segment directly above?
4. What is the relationship of that segment to the segment directly below?
5. What is the relationship of that segment to the plumb line?
6. Begin to imagine possible symptoms or complaints that this person might have.
7. Compare your thoughts with client's interest or complaint.

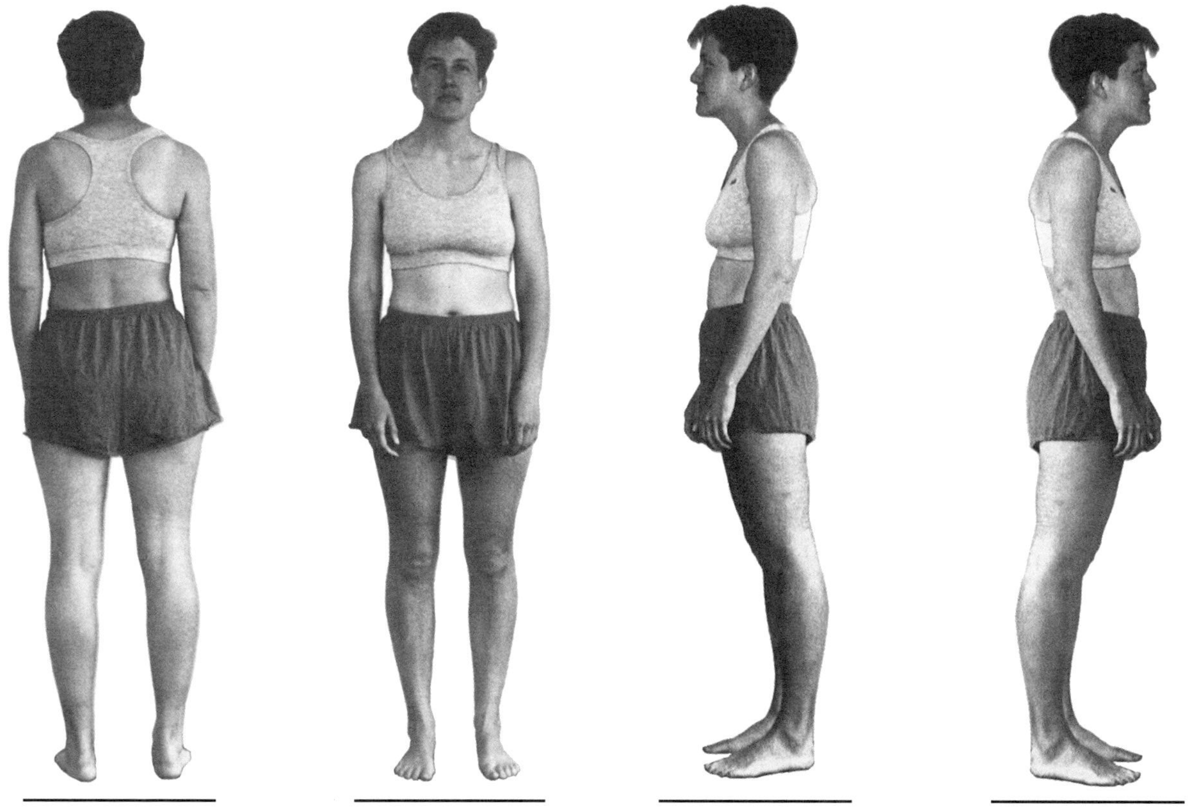

A B C D

Identifying and Selecting the Main Track

As you begin to prioritize which view—with its corresponding segment and relationship—gives you the most information, you will select the appropriate symbols for:

1. anterior-to-posterior placement
2. up-to-down placement
3. right or left placement
4. rotations

As you identify which relationship gives you the most information, stay with that same track of comparing:

1. rotation to rotation
2. placement (displacement right with displacement left)
3. tilts (right shoulder high to left shoulder low, and right hip low to left hip high)

For now, it can be confusing to mix tracks, as in comparing a left hip rotation with a high right shoulder placement, and an upper chest centered posteriorly to the pelvis.

Exercise 2.35

Now practice by notating each view.

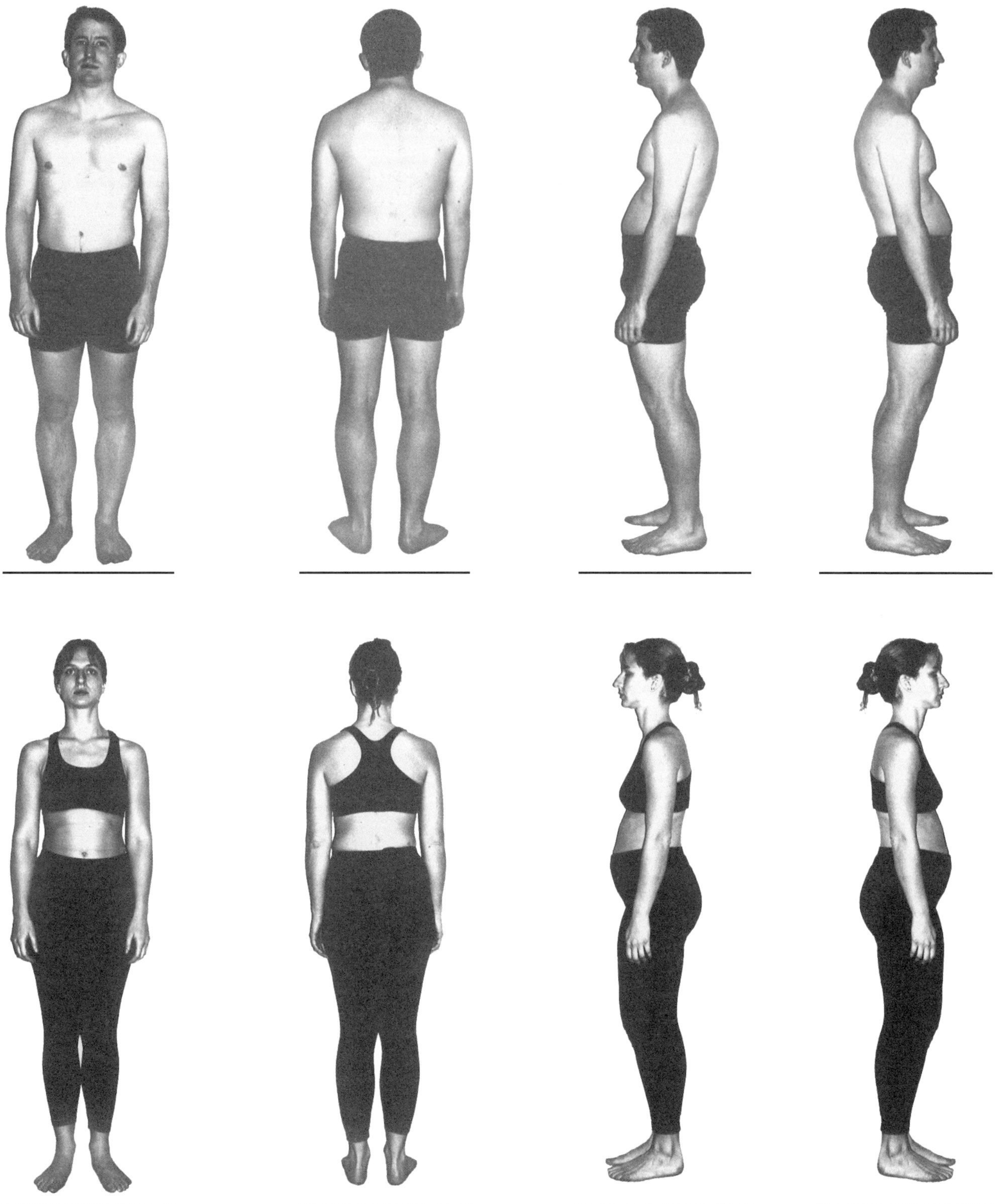

The Alignment of Extremities

Placement of the Arms

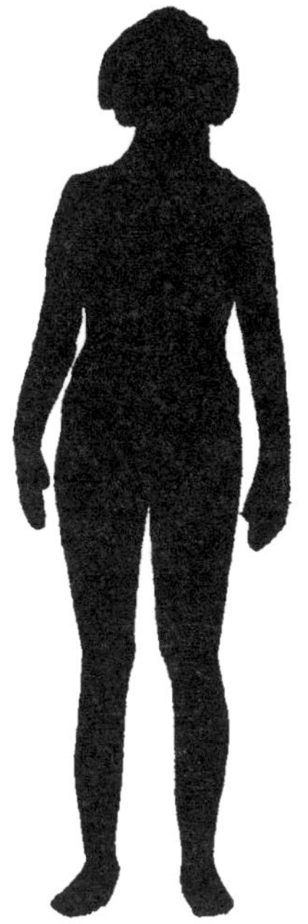

An easy way to begin assessing the arm position is to look for the contrast between the contour of the trunk and arms and the space in-between them.

What is the body pattern that matches or causes the arm position? We could describe the right arm position as being closer to the body and internally rotated at the shoulder. The left arm is farther away from the body.

The space between the arms and the body can give you much information about the alignment of the trunk. Sometimes one arm will appear closer to the body than the other. This might indicate a shift or rotation in the pelvis or shoulders. Look at the following photos with your focus on the size and shape of the space between the arms and the trunk. Note the information this gives you about the alignment of the trunk.

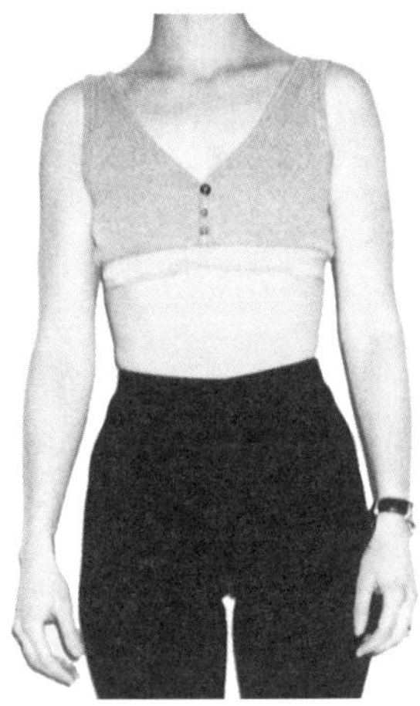

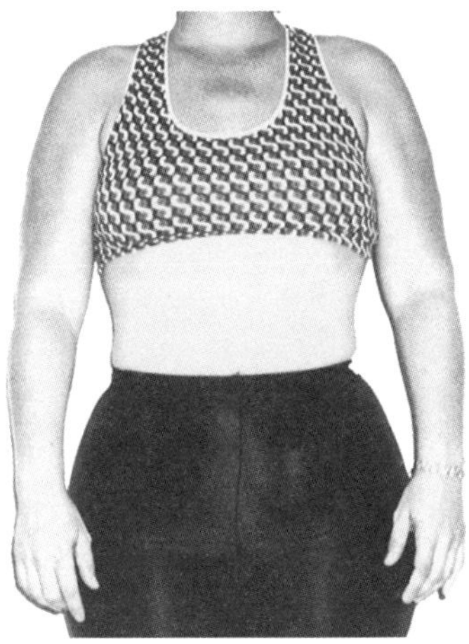

Exercise 2.36

Following the previous observation sequence, continue to check the placement of the segments within the three planes.

1. the anterior to posterior relationship/sagittal plane
2. the right-to-left or side-to-side relationship/frontal plane
3. the up-to-down relationship/transverse plane

For the following figures and exercises:

A. Observe

1. The relationship of one arm to the other or to the trunk

    a. the transverse plane—up to down placement

    b. the frontal plane—abduction/adduction (out from or in toward the midline)

    c. the sagittal plane—anterior or posterior placement (including flexion and extension of the joints)

    d. internal and external rotation at the joints (which can include pronation and supination of the forearms)

    e. a combination of the above

    f. the amount and shape of space between the arm and trunk

2. The relationship of one segment to another (upper to lower arm or hand)

    a. the frontal plane—abduction/adduction (out from or in toward the midline)

    b. the sagittal plane—anterior to posterior placement (flexion and extension)

    c. internal and external rotation at the joints (pronation or supination)

    d. a combination of the above

B. Notate

Exercise 2.37

The following are three of the arm's basic positions.

1. abducted/adducted

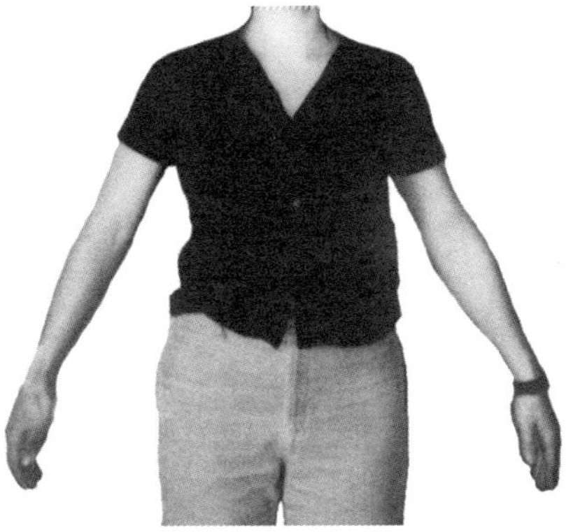

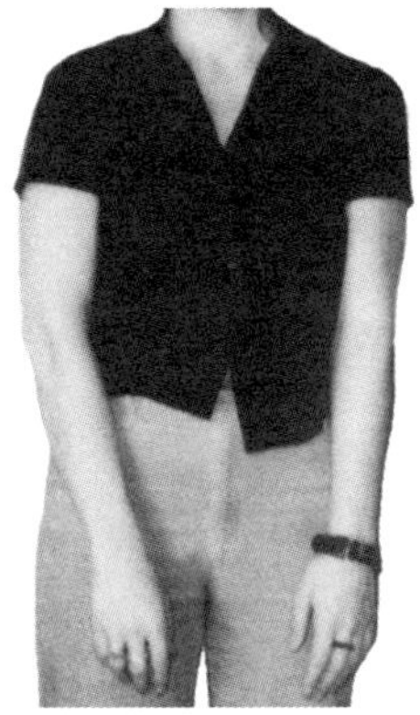

2. anterior/posterior placement (flexion or extension)

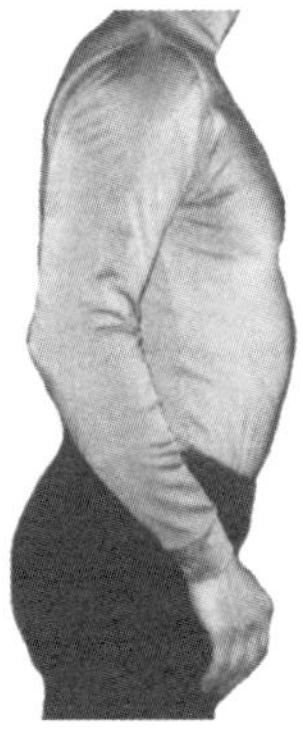

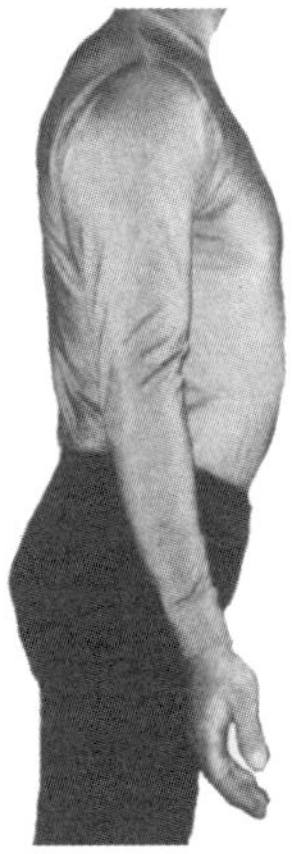

3. interior/exterior rotation

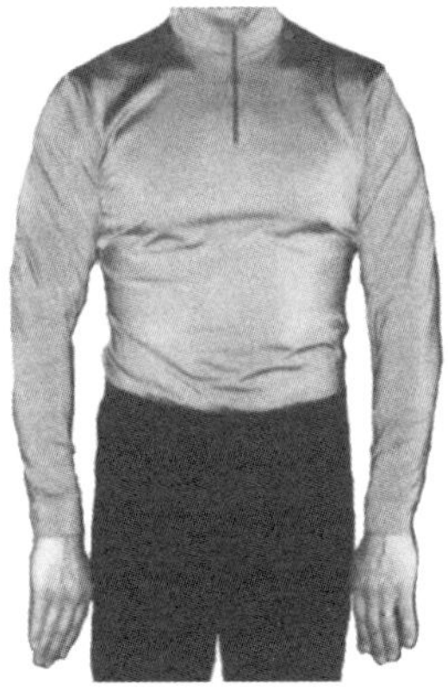

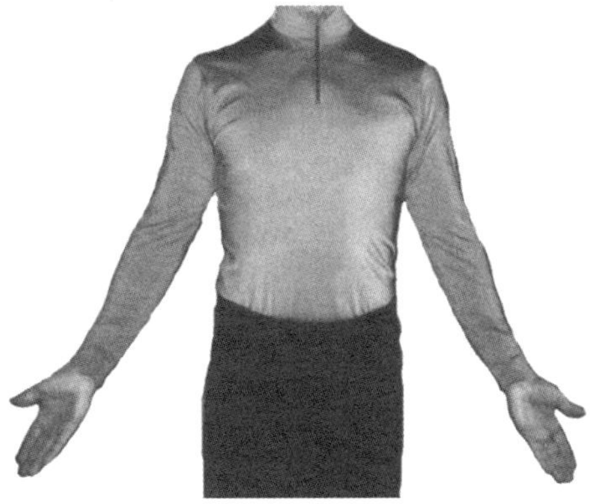

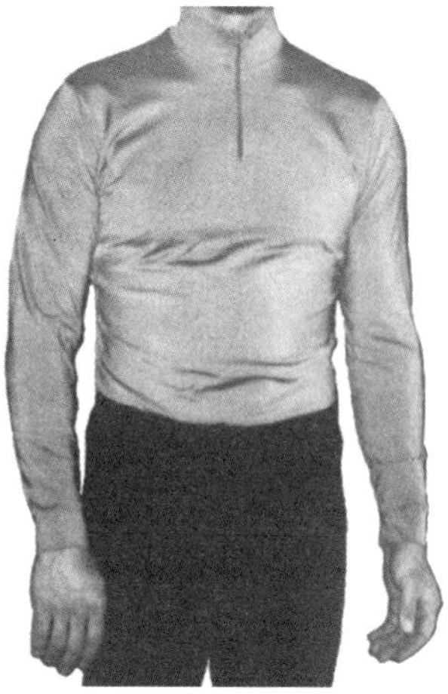

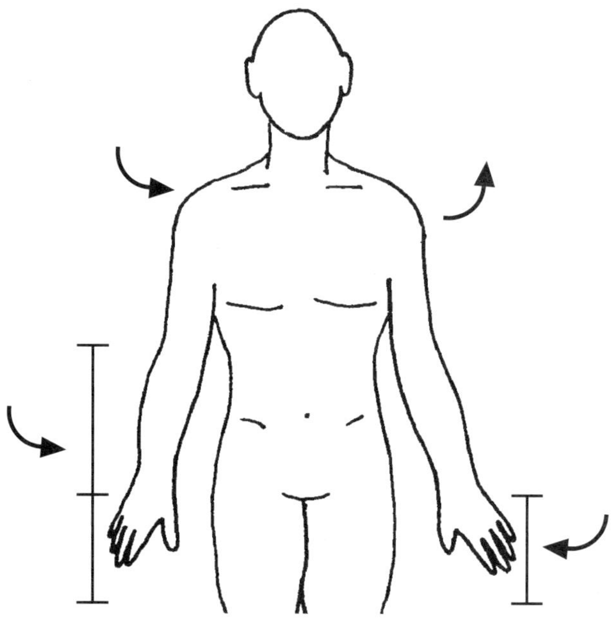

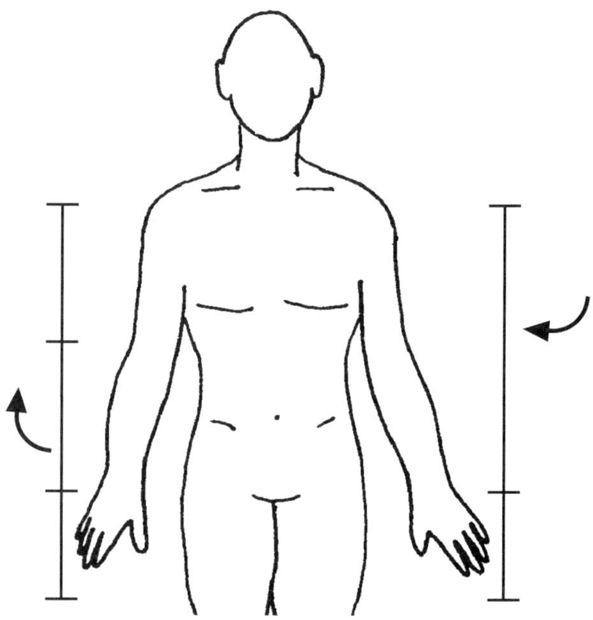

Observe the placement of the arm toward the front or back of the body. If one arm seems more forward, it may mean that the whole side, including the trunk, is in the same rotation.

Look at the following photos and notice that the segments of the right and left arms are placed quite differently.

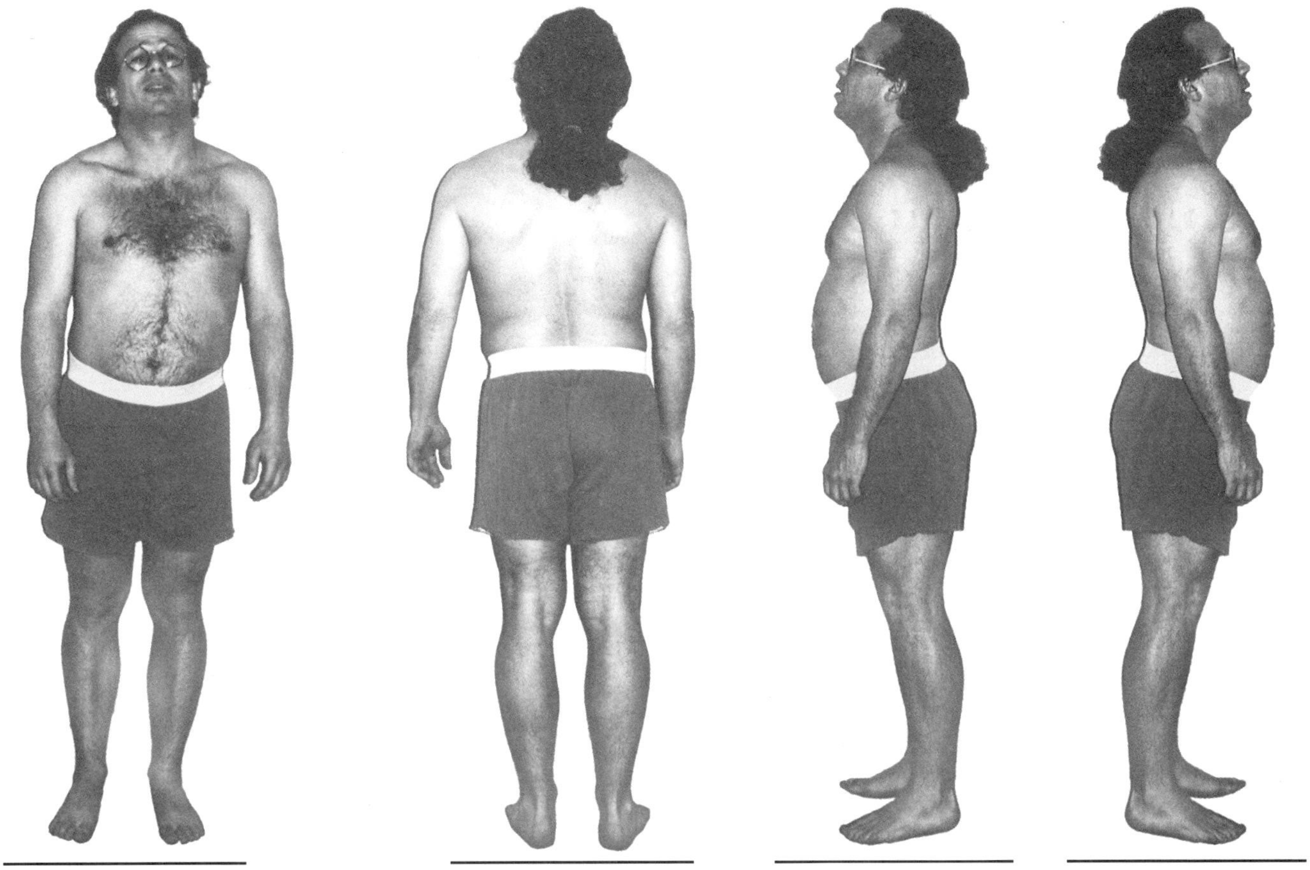

These anterior and posterior views demonstrate that the left arm is forward of the plumb line (displaced anteriorly) and the right arm is behind it.

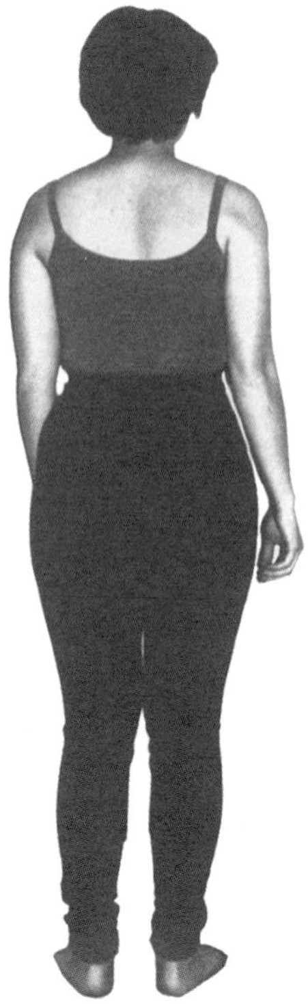

From the side view, you can confirm that the woman's right upper chest is behind the plumb line (displaced posteriorly), which also places her right arm behind it.

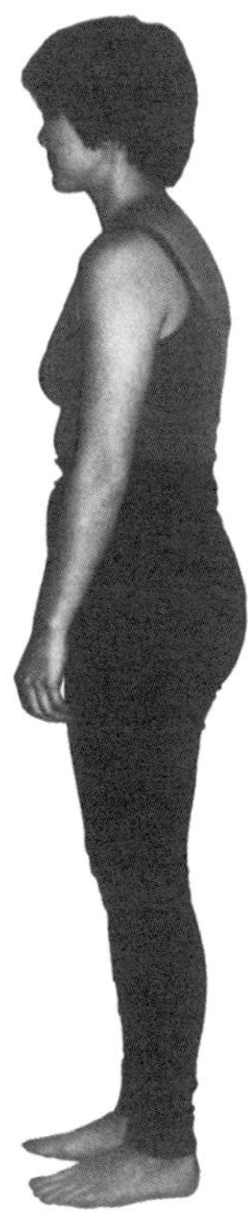

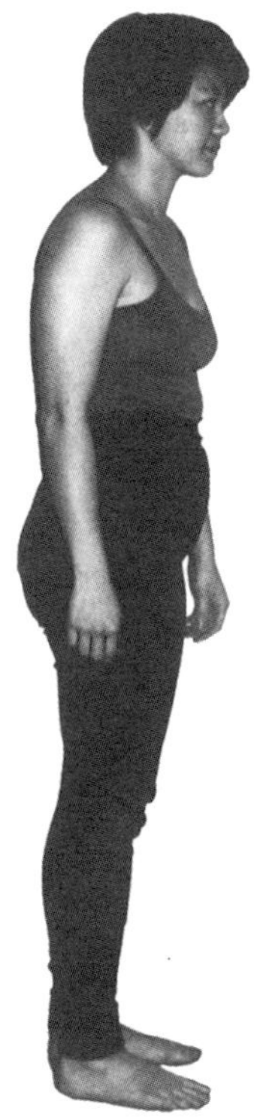

Now, look at these photos. Notice that although the woman's right upper and lower chest are behind the plumb line (displaced posteriorly), her right arm is internally rotated (pronated), which brings her forearm forward (anterior).

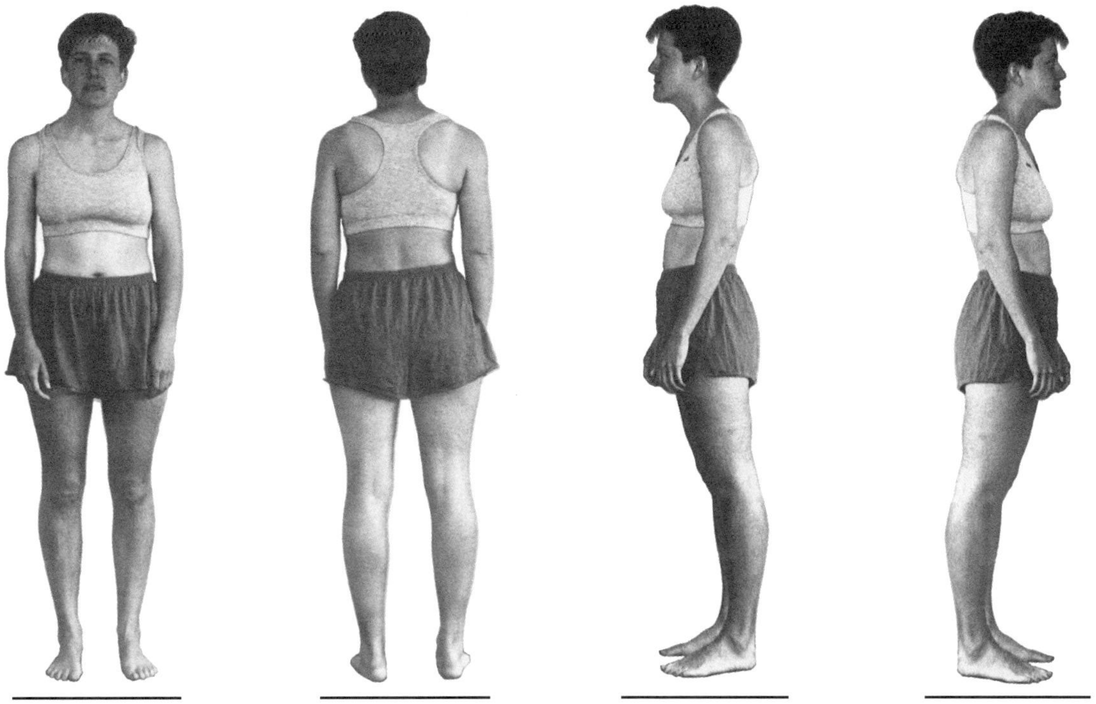

More specifically, notice that her left forearm is more internally rotated to counterbalance her chest rotation in an attempt to centralize her position.

In the front and back views, we could select several tracks: displacement right and left, up to down, or rotation. As one example, let us select the rotation track. Some of the rotations would be:

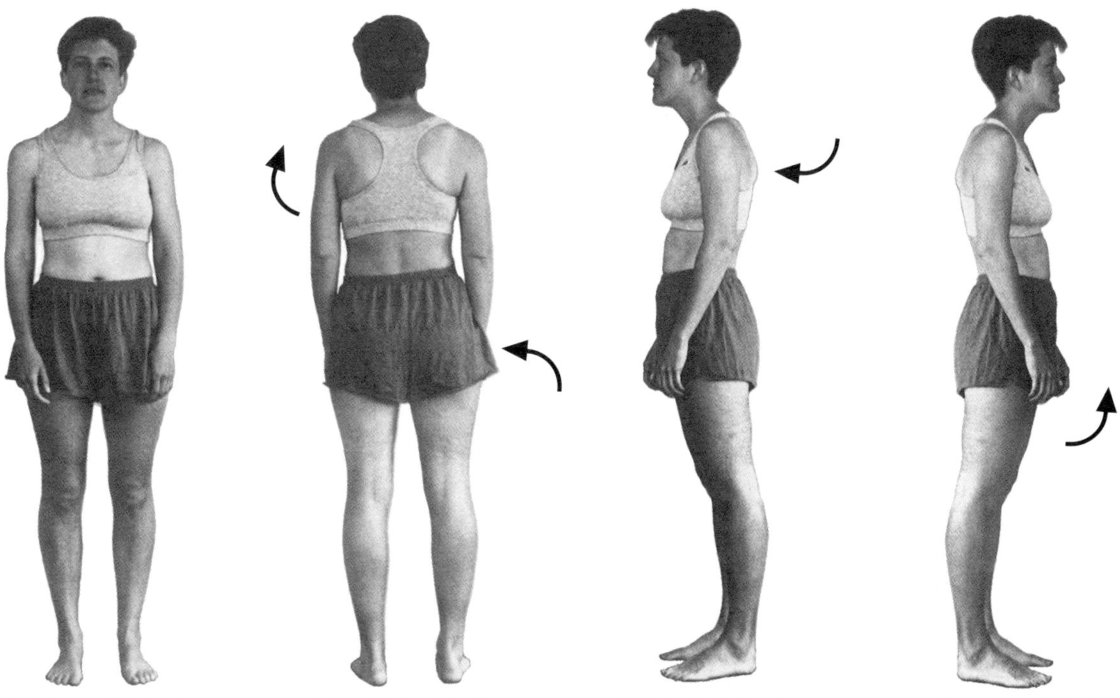

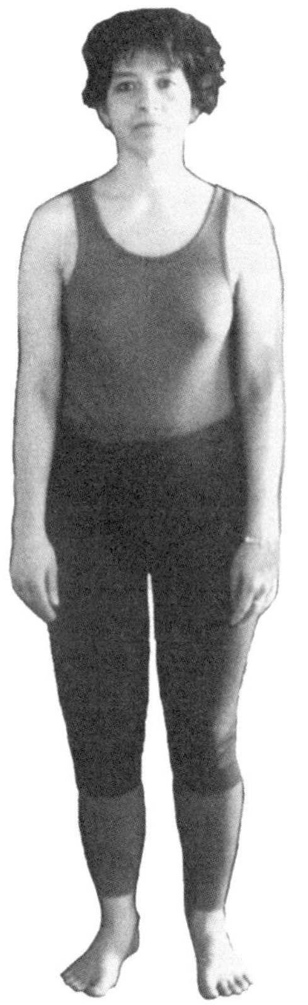

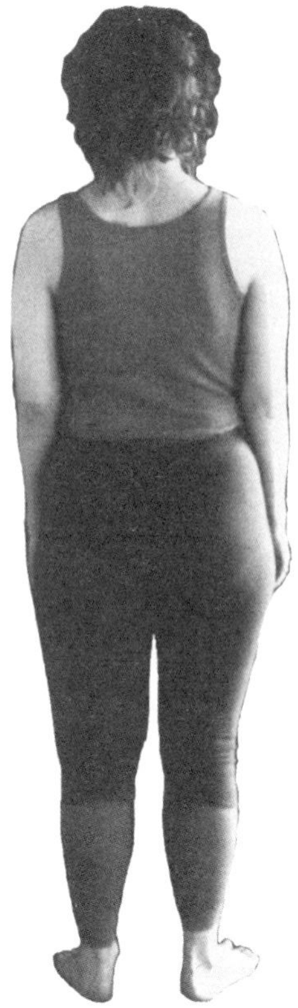

Her left shoulder appears more central, if not slightly externally rotated, while her right shoulder is more internally rotated, which makes her right forearm even more internally rotated.

Her weight, particularly her chest, seems shifted to the left, which puts her right arm farther away from the center. This contributes to the amount of rotation necessary for her right arm to approach the center.

Notation Combinations

The following examples indicate an easy way to notate the arm position. We are able to see the relationship of the arm to the plumb line, as well as the relationships within the arm.

In this example, you can see that the woman's arms, at the shoulders, and the forearms need to go forward and internally rotate, in order to counterbalance her chest displacement posteriorly.

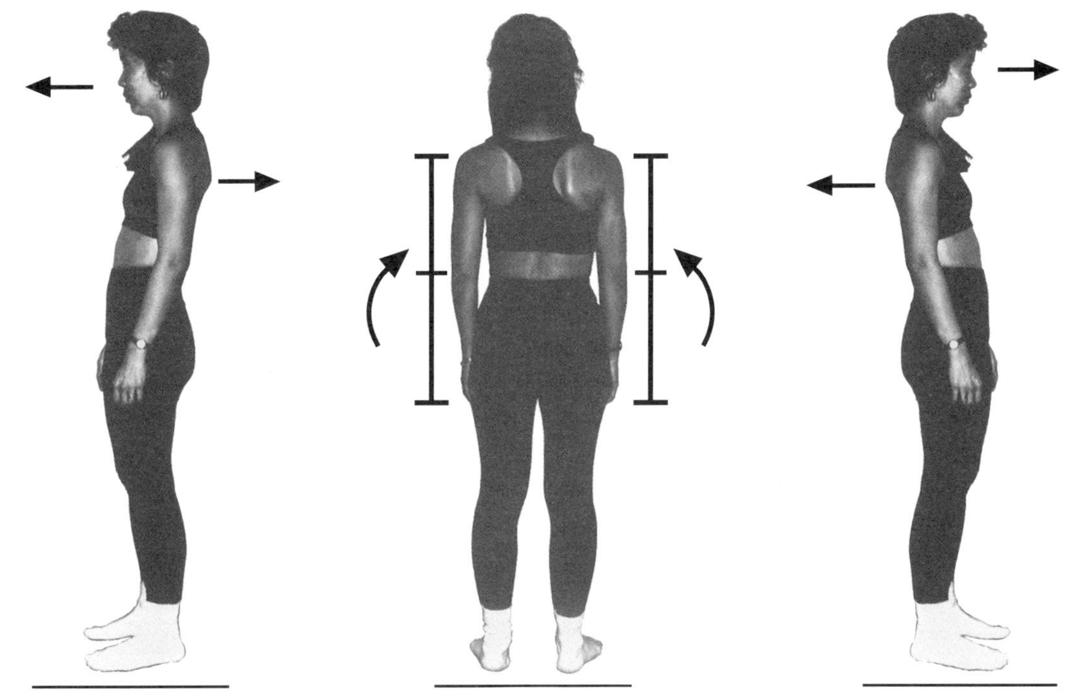

The client in the following photo also demonstrates the counterbalance of the arm placed anteriorly (forward of the lateral midline) in order to balance the chest, which is posteriorly displaced (behind the plumb line).

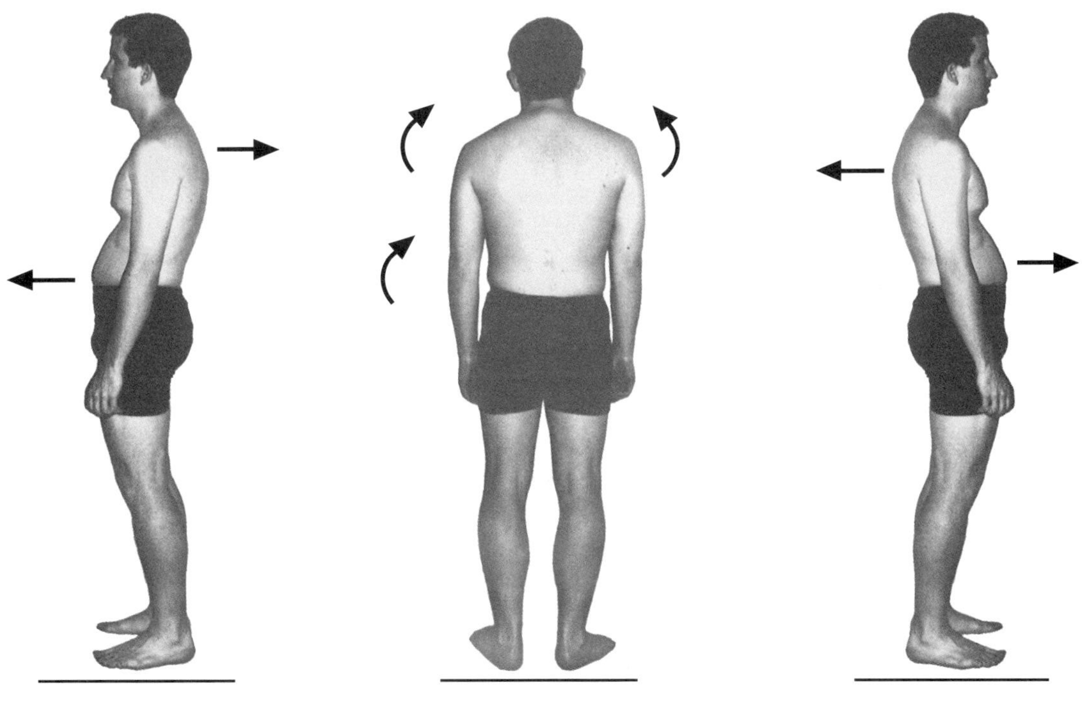

Legs

For the following figures and exercises, return to the illustrations on p. 111.

A. Observe

1. The relationship of one leg to the other
   - a. the transverse plane–up to down placement
   - b. the frontal plane––abduction/adduction (out from or in toward the midline)
   - c. the sagittal plane–anterior or posterior placement (including flexion and extension)
   - d. internal and external rotation at the joints (usually can be observed from the front and back views as well as from the sides)
   - e. combination of the above
   - f. the amount and shape of the space between them
2. The relationship of one segment to another (upper to lower leg, as well as to the foot)
   - a. the frontal plane––abduction/adduction (out from or in toward the midline)
   - b. the sagittal plane–anterior or posterior placement (including flexion and extension)
   - c. internal and external rotation at the joints (usually can be observed from the front and back views as well as from the sides)
   - d. a combination of the above

B. Notate

Legs—Examples of Placement and Rotation

There are three basic positions of the legs.

1. abducted/adducted

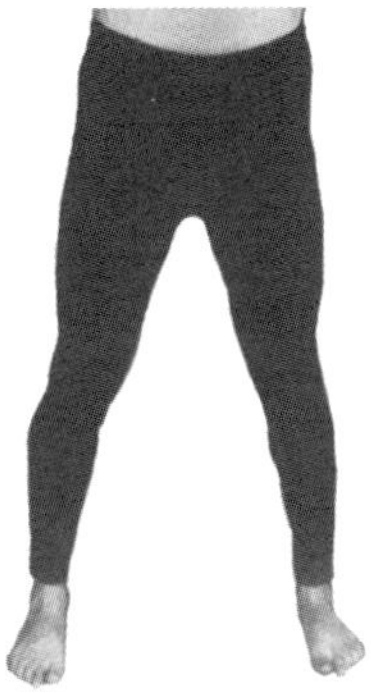

2. anterior/posterior placement (examples of knee flexion and hyperextension)

3. internal/external rotation

4. a combination of the above.

You may want to use the earlier exercise of outlining the segments, marking the segmental centers, and connecting the centers as a reference before you add rotations.

Exercise 2.38

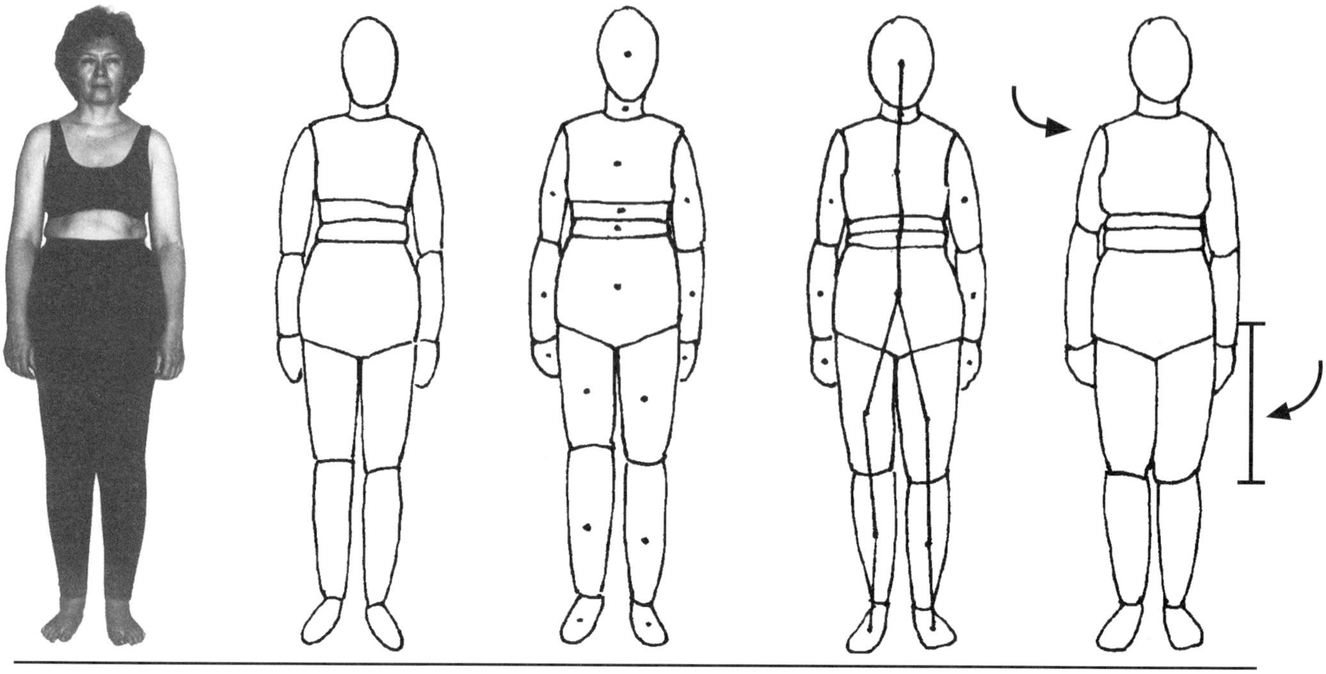

Now, look at the back view of this person's pattern. Does it confirm what you see in the front view? If not, see if you can identify whether the different appearance is due to rotation.

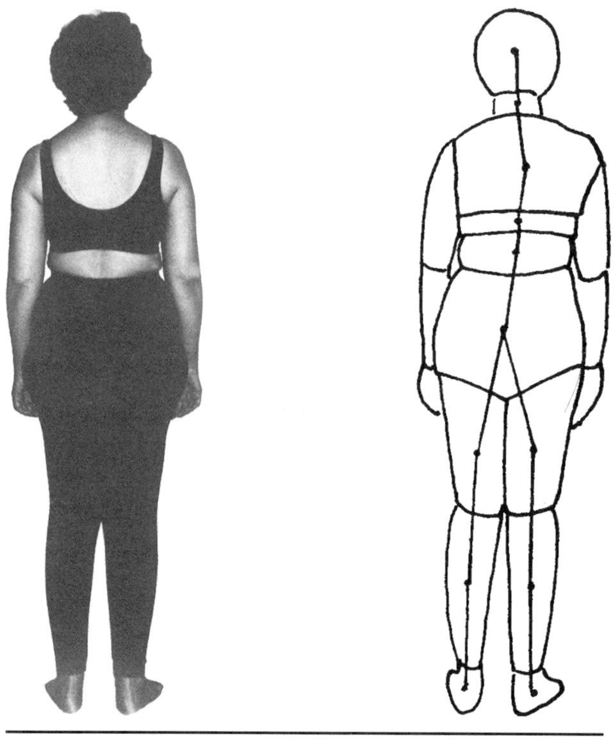

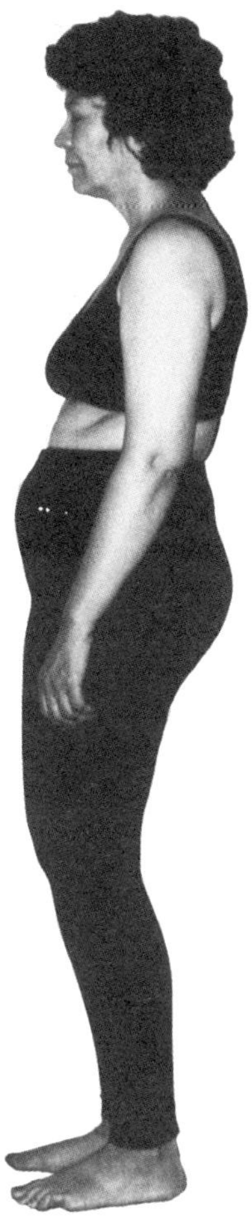 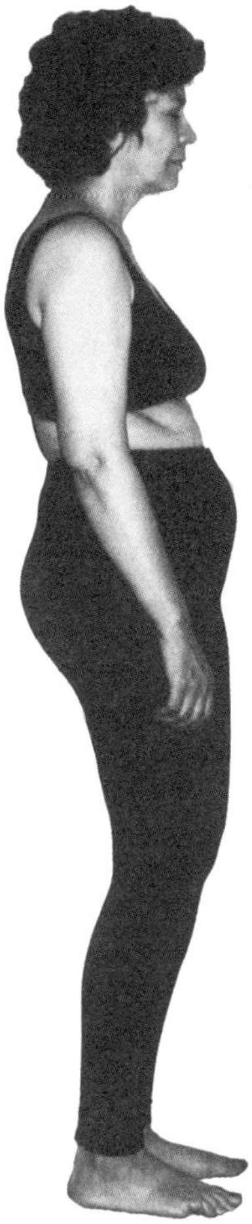

Now compare the side view with the front and back views of this photo on the previous page. Do you see that you can begin to hypothesize certain patterns in the other views before you even observe them?

People sometimes adopt certain standing positions either out of habit, because of physical need, or sometimes they stand in a way that they have been told is good for them.

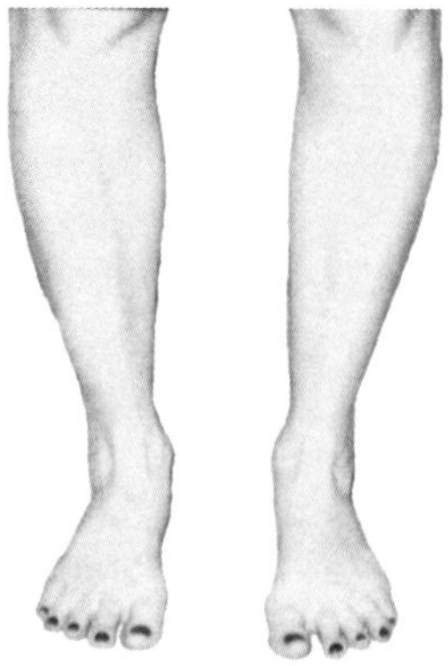

Feet facing straight forward

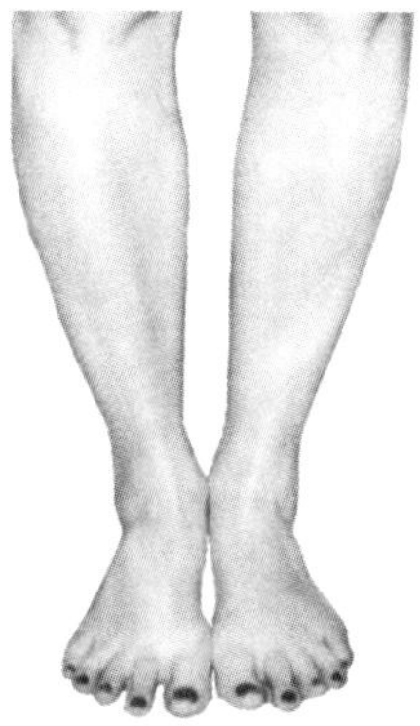

Feet close together

Knees bent

Midtest

Look again at this photo, which you used for the pretest. Note what you see in terms of alignment, tilts, and rotations.

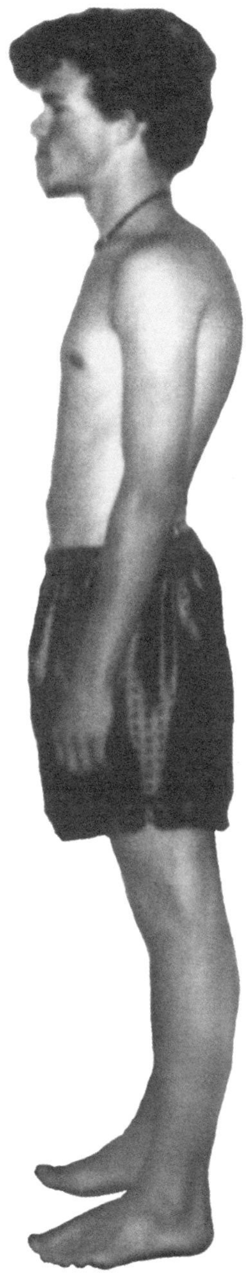

Are your observations the same? Are you gathering different information or seeing additional relationships?

CHAPTER

3

Aston-Mechanics℠

Introduction

For many years, I taught from a more standard or familiar postural model. After 15 years of experience using the standard model, I began to correlate my observations.

I rarely had seen:

1. Ideal posture in standing position or while walking.
2. Symmetrical alignment of the right and left sides of the body.
3. Even distribution of the body weight to both feet.
4. Sustenance of good posture with ease.

On the rare occasion that I observed these favorable conditions, they were associated with a person who performed them with concentrated effort.

I particularly witnessed this holding of so-called ideal or good posture when I taught movement education. What became obvious to me was the sense that, instead of my students showing me that they were 20 people in the room with 20 different examples of good posture, they exemplified only one idea of so-called ideal posture. I was puzzled by this loss of individuality, and I realized that I did not want to encourage this pattern.

As I began to explore postural models, as well as their origins and influences on furniture, shoes, and other products, I learned that postural ideas dictate our body use and, therefore, our body shapes.

As I explored different options in teaching movement, I discovered:

1. That the more I developed the skill to see the details of the whole body pattern, the more success I experienced in problem solving.
2. The importance of teaching movement of the body in smaller increments, in order to help the person learn more easily how to achieve change that is long lasting. This is a better alternative than imposing an ideal model that is beyond a person's ability to use.
3. That the asymmetry of the deep muscles (heart, liver, etc.) has a significant influence on the more superficial muscles, and creates slight asymmetries throughout the body.

4. That trying to contain these influences would create an unnecessary holding pattern, negatively influencing the function of internal organs which, in turn, affect the vitality of the human organism.
5. The importance of seeing the three-dimensional internal volume of:
    a. One segment to itself (the abdomen looks expanded)
    b. One segment in relation to other segments (the back looks short while the anterior chest also looks short)
    c. One segment in relation to the whole body pattern
    d. How three areas might play a part in creating symptoms such as those of low back pain
6. That the ability to see not only the alignment of body segments, but also their shape and dimension, provides much information on their function.
    a. An expanded chest may increase breath but decrease shock absorption through the spine.
    b. A diminished volume might decrease breath and increase stress on the low back.
7. How to see the whole body pattern, instead of one isolated symptom or area, in order to provide information on what secondary areas may be contributing to the primary area.

My constant attention to seeing alignment and dimensional relationships revealed the need for a different postural model that would not restrict dimension or function, whether in stationary balance or in action.

This discovery created a challenge for me, as it posed quite a contrast to my previous learning. I was looking for the best structural position that would encourage a fully proportional dynamic body in function, as well as a functional model that would reinforce and improve the structure. As a result of my quest, a new paradigm emerged.

Stationary Balance

The following structural guideline emerged from my observations, as well as the feedback I received from people who reported feeling more balanced as a result of these ideas. This ease of balance resulted when no one body segment was borrowed from, compressed, or restricted by any other segment, in order to hold so-called good posture.

Side View

1. The central line of the body is inclined forward 2–5 degrees of perpendicular to the floor.
2. The center of mass moves 2–5 degrees in all directions in relation to the apex of the longitudinal arch of the foot.
3. The angle of the legs is externally rotated 5–15 degrees at the hip joint, creating a 5–15-degree turn outward of the feet.
4. The knees face slightly outward, matching the open angle of the hip and ankles.
5. In the frontal view, the width of the foot placement aligns the heels more directly under the line of the ischial tuberosities and hip joint. It also aligns the forefoot under the line of the greater trochanters and shoulder joint.
6. The shoulder joint is aligned directly over the hip joint.

7. The shoulder joint is slightly externally rotated in order to match the external rotation of the hip joint and the body's center of gravity, which is 2–5 degrees forward.
8. The body's forward incline of 2–5 degrees maintains the extension curve of the cervical spine.
9. The head rests on the spine slightly angled forward and down 2+ degrees congruent with the lean of the body.

This position in stationary balance is not static but dynamic. The range of motion (ROM) is 2–5 degrees in all planes. This guideline describes a possibility for optimum balance. The slight changes in degrees also will change the amount of flexion/extension and internal/external rotation, as well as the forward/backward position of all segments.

These references can be used as a guideline for seeing the body in stationary balance. However, because the body is always in motion and every body is unique, the reference point is used for comparison and is not intended to be a model imposed on every structure or person.

In fact, some people will need to have a stance in which the legs are internally rotated to match an anteversion of the hip joint. This stance may be needed for a long-term or perhaps a temporary stance for overall balance until the person can be assisted toward continued change.

Aston-Mechanics[SM]

Many therapists have found the principles of Aston-Patterning to be most helpful in working with their patients. This section explains the five basic Aston Principles for your learning and exploration.

Aston-Mechanics is a new math for body usage in action and at rest, and for ergonomic product design. It offers a guideline for body posture. This perspective can influence every action, reaction to forces (gravity and GRF), and interaction (with objects as in sports, products as in ergonomics, and with people).

Five Aston Principles

Please remember that the following principles serve as a guideline with numerous possibilities and are not to be models imposed on anyone.

Familiar models of posture teach us that the right and left sides of the body are to be symmetrical.

Principle 1

The body is supposed to be slightly asymmetrical from one side (right) to the other (left).

Rationale

1. This is consistent with the asymmetry of the internal structures, such as the size, weight, and placement of one heart and one liver, and the fact that the lungs have three lobes on the right and two on the left.
2. The body structurally records and asymetrically compensates for trauma, injuries, surgeries, and so forth.
3. The asymmetry increases as the body compensates for the conditions described in statements No. 1 and No. 2, as well as for additional habitual body-usage patterns.

Benefit

The asymmetry principle is extremely important with respect to movement. In natural objects just as in the body, one observes asymmetry within the strata of the Earth, the shape of shells, and the rings in the cross-section of a tree. Likewise, in movement, there is constant asymmetry and non-linear designs. Where the aspects of density, tempos, and direction interact, we see manifestations such as waves, a snowfall, and a flowing river move in non-linear patterns. The body is the same. It is made up of a continuum of many different densities from fluids to solids. Also, it is made up of different tempos, in the functions of breath and circulation. These, combined with our structural asymmetry, dictate that we also need to move in non-linear ways. I am not encouraging that anyone exaggerate asymmetry. I am proposing that, because of injuries, history, or viscera, the body becomes asymmetrical.

We need not be alarmed by the slight asymmetries we see in the body. Balance is the negotiation of asymmetrical differences within a given structure. We merely need to teach the client to use both sides (right and left) uniquely. Being able to *see* enables us to investigate what is a necessary or unnecessary asymmetrical pattern to the many actions, ADLs, sports, work, and so forth.

From the side view of (A), we see that the body's segments in familiar postural models are aligned around a line at 0 degrees—perpendicular to the floor. The body (B) is inclined at approximately 2 degrees forward.

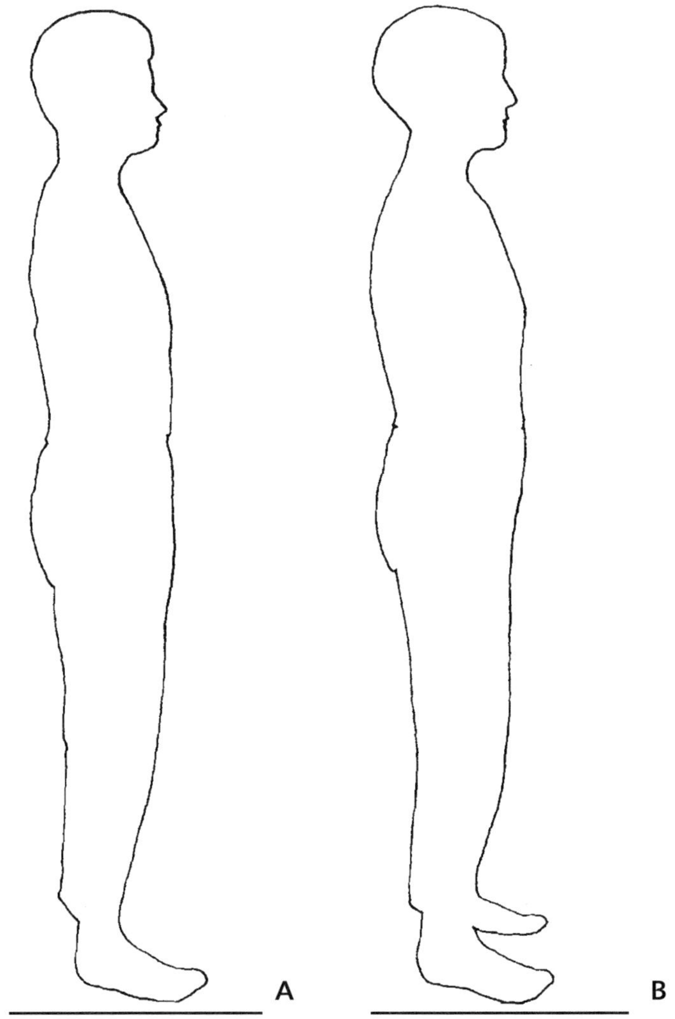

Principle 2

The alignment of the body segments above the foot will angle approximately 2–5 degrees forward of the vertical 0 degree.

Rationale

The Aston position puts the weight bearing through all the centers of the joints but aligns the center of each segment slightly anterior to the segment below, in order to be aligned over the whole foot. This poses a contrast to standard alignment theory, which uses the 0 degree/perpendicular line to determine the weight bearing as either anterior or posterior to the joints, and the medical model, which places weight bearing in the center of each joint on the 0-degree line.

Benefit

This slight forward inclination allows more even distribution of weight and effort through the anterior and posterior aspects of the body. This position also optimizes the forces of gravity and GRF to translate more evenly throughout the spinal curves, creating more of a lengthening effect upon impact.

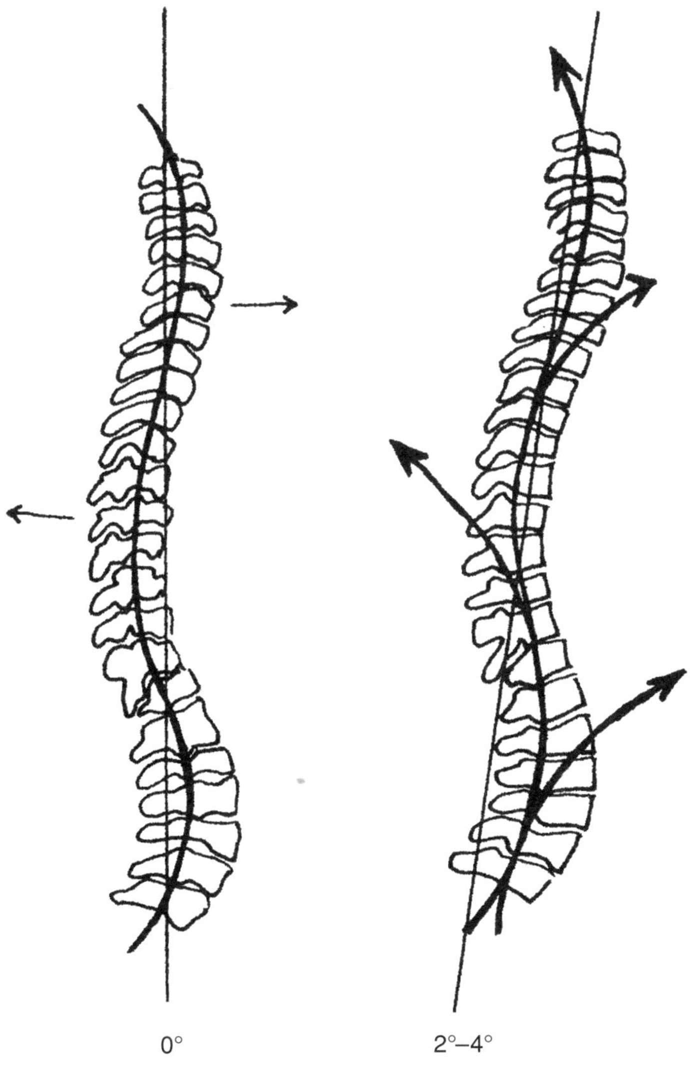

Traditionally, the body aligns itself around a central axis, which creates the midline of the body.

Principle 3

Because the body constantly is moving, the midline is a range across an area made up of six borders. These landmarks are:

1. Right side of sternum down through the pubic symphysis.
2. Left side of sternum down through the pubic symphysis.
3. Left side of spine down through Sacrum Level 1.
4. Right side of spine down through Sacrum Level 1.
5. Posterior aspect of hip joint up to shoulder joint.
6. Anterior aspect of hip joint up to shoulder joint.

Rationale

In stationary balance, the body continuously sways slightly to the right and left, forward and backward, up and down, in order to negotiate with the gravitational and GRF forces. I call this dynamic balance the range of neutral (RON).

Benefit

This movement aligns the body alternately over each side, causing a loading and unloading or alternating compression and distraction. This movement from side to side allows one side to rest while the other side is more active (under stress).

This compression and distraction (loading and unloading) of body weight creates a massage-like action within the body.

The muscles, organs, circulatory system, breathing and, we hope, mental activities constantly are encouraging us to move in one direction or another. All of these movements require that we find a new balance from one moment to the next. The plumb-line bob moves.

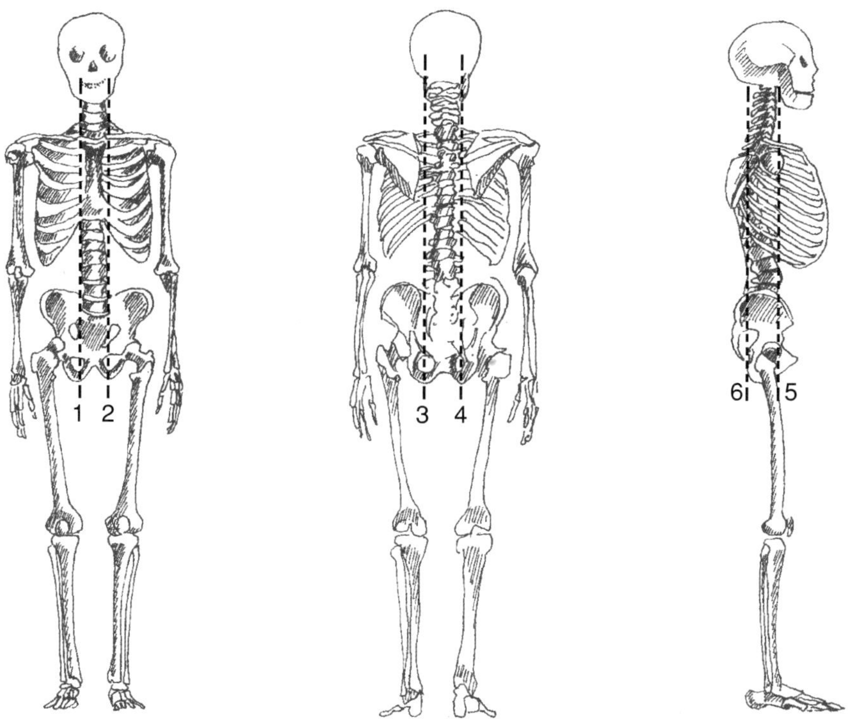

From the front view in familiar models, the neutral position of the body places the feet facing straight forward or 2–4 degrees in external rotation.

Principle 4

The position of the foot is consistent with the open position of the hip and leg, which ranges from 5–15 degrees of external rotation. Please note that the angle is determined by the individual's current body options or restrictions.

Rationale

The open stance gives skeletal support to the depth and width of the body segments above it. The slight external rotation places the calcaneus in alignment with the ischial tuberosities and, therefore, gives support to the body's midline. The position and angle of the forefoot support the hip joint, shoulder joint, and arms.

Benefit

The position allows the weight and effort to be more evenly distributed throughout the skeleton and soft tissue. The joints are more evenly loaded, which maximizes more optimal joint ROM.

In familiar postural models, the side view of the body's neutral alignment is centered around a perpendicular line that runs through, or slightly anterior to, the lateral malleolus.

Principle 5

The plumb line falls slightly anterior to the lateral malleolus on a foot that is slightly externally rotated 5–15 degrees, putting the weight more on the whole foot.

Remember that the plumb line would be in continuous motion in order to accommodate the principle of RON.

Rationale

This neutral alignment distributes weight onto the whole foot and is centered close to the apex of the longitudinal arch.

The continuous movement or oscillation allows the weight bearing to be viewed not as a single point but as an area.

In standing balance, the foot is designed to be the greatest weight-bearing segment. Centering the body over the entire foot in order to use its whole length creates the need to move the axis point to the front of the ankle joint, where it hinges, and move the plumb line from 2–5 degrees, forward and then backward.

Benefit

The moving plumb line increases the use of the whole foot—its length, depth, and width—which provides the fullest base of support (BOS).

This alignment also increases the ease of using the forces of gravity and GRF more directly through the spring-action arch of the foot and to the whole body. This poses a contrast to the weight being centered more posteriorly on the foot (heel), which usually encourages increased tension or compression in the effort to sustain such an alignment.

Even in the standing position, the BOS continually is moving. This dynamic posture encourages the body segments above the feet to negotiate the constant changes in position. These micro-movements enhance blood flow, exchange of nutrients, and joint lubrication, all of which are necessary for healthy tissue.

Alignment of the body centered over the malleoli tends to direct weight bearing onto the posterior third of the foot, and posterior to the midpoint of each joint above. When the weight is centered slightly in front of the ankle, several things occur.

1. Weight bearing occurs more directly through the entire foot and is supported through the centers of each joint above it.
2. Standing balance becomes a dynamic phenomenon whose center shifts from right foot to left foot and mid-foot to forward or backward, in order to place an even distribution of weight over the entire foot. The whole foot is used as a BOS.

Neutral

Neutral is a specific configuration of alignment and dimension at any particular moment in time during which the body segments are in their best available relationship. Neutral is not a point but an area that is a three dimensionally balanced posture, and it is in continuous motion. This dynamic neutral, or RON, is from right to left, front to back, and down to up. These directions are combined to accommodate the internal structures, breath, thought, and so forth.

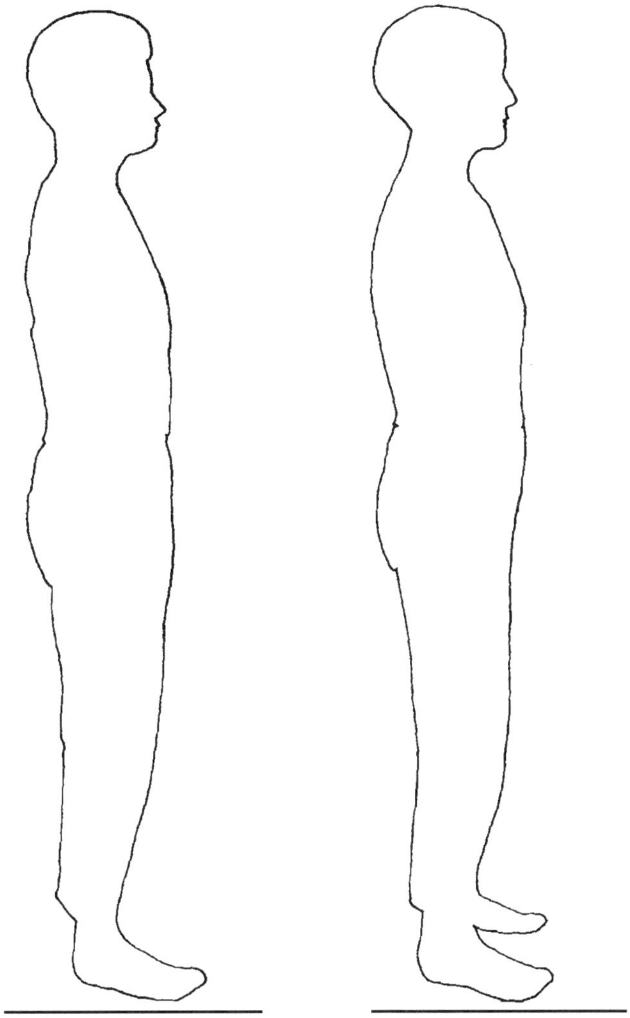

0–2° **4–5°**

Range of Neutral from the Front and Back Views

Demo

The midline can be drawn from (right-left borders) anterior and posterior, as in Principle No. 3.

Back View

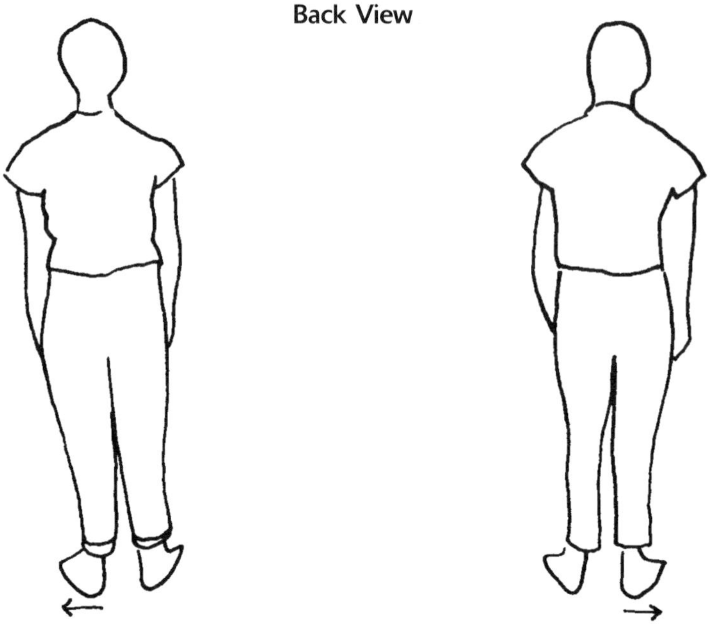

Range over left foot **Range over right foot**

Front View

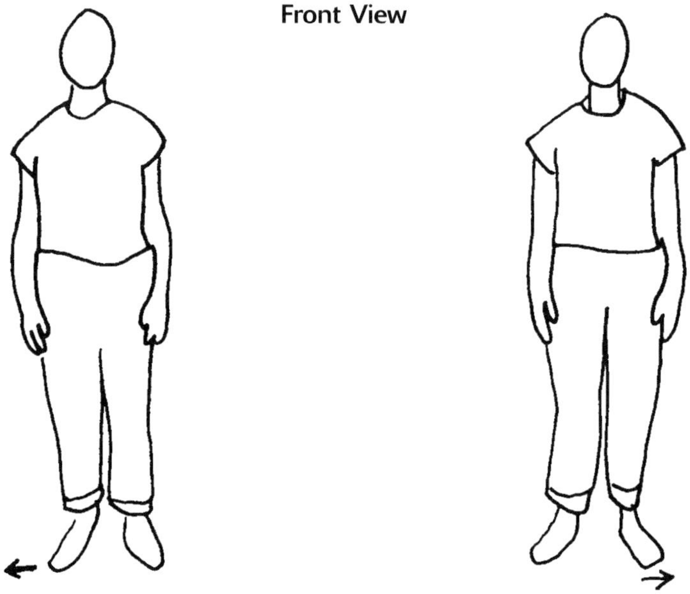

Although the person is in a slightly different position, you still can use either of these illustrations to draw your RON and to compare placement and dimensional relationships.

The sway that places a person's center of gravity closer to the left or right side means that his or her plumb line, which usually is vertical, would need to be on a diagonal. For that reason, let us refer to it as the midline range.

Range of Neutral (RON) for the Leg Position

External rotation of the leg causes 5–15 degrees of external rotation in the foot.

> Please note: The RON of 5 15 degrees accommodates the unique body patterns of the male and female pelvis, as well as specific asymmetrical body patterns. Additionally, the two legs may need different degrees of rotation, including internal rotation, in order to negotiate a limited range.

Legs

The foot position is determined by the amount of range of motion at the hip joint. Each leg needs to support its own side of the body. If the placement of the foot is wider than the same hip joint, it reduces the support for the medial portion of that joint. Likewise, if the feet are placed too close together, the lateral portion of the hip joint has less support. We find that the optimal position needed to support the full width of each side is for the feet to be slightly turned out. The heels are closer to the midline to support the midline structures. The front of the foot is placed slightly wider to support the lateral line.

The neutral alignment of the legs, in the view of this paradigm, is for the feet to be aligned with the heels under the ischial tuberosities. This poses a contrast to some postural models that place the feet in a narrow stance a few inches apart or in one that is shoulder-width apart. Because the two sides of the body are sometimes quite asymmetrical, they need an asymmetrical position of the feet in order to feel more comfortable and supported.

Position of Feet

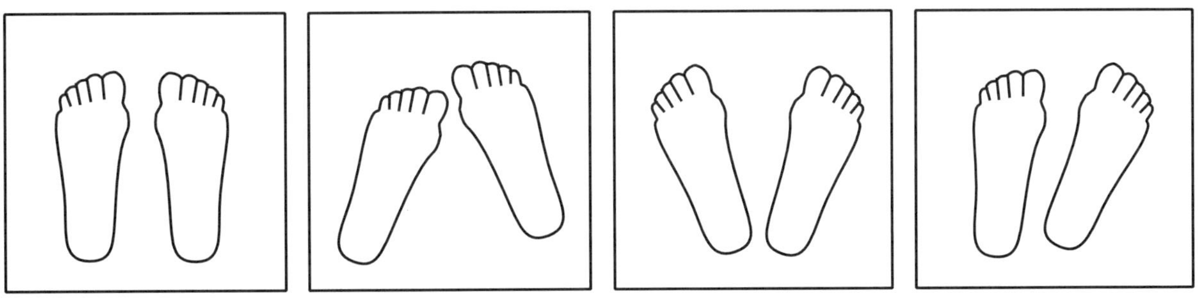

Straight   In   Out   Asymmetrical

Note: Neutral does not look the same on all people. It is a mathematical equation specific to each person and to each moment, because a mere thought can change neutral posture. Neutral is when the body does its best to integrate all the history of influences (e.g., genetics, injury, sports, beliefs, personality, emotion) and can rest between the forces of gravity and GRF in postures that minimize the need for any segments to overwork or restrict other segments. Neutral is available to the person who has an extremely flexed posture from osteoarthritis. The person in a wheelchair also has an available RON.

To help people find their RON, factor in the greatest limitations. One patient might need to use crutches for six weeks. A stroke patient might not have much use of his or her right side. Another may be grieving over the loss of a loved one. Yet another might have a strong scoliosis. These (primary focus) limitations can create specific demands on the body (secondary focus).

The significance of looking at the primary and secondary foci is to avoid the negative consequences of assisting the secondary areas in changing beyond the capability of the primary limitation.

Seeing Dimension

So far, we have been studying the position, or alignment, of the body segments. Now let us observe the body's shape or dimension.

Dimension

This refers to the internal volume of any given part and is one method of observing the body. The three dimensions for observing are length, width, and depth. Together they comprise the internal volume that gives us a view of the space inside a segment.

This perspective of observing body dimension gives us very different information than that derived from looking at the outline or landmarks of the body.

A body can appear to be in good posture, with its key landmarks in line; however, if the person has decreased dimension in the anterior chest and posterior mid-trunk, he or she could experience increased compression and shear forces in the low back.

We begin to understand that in order for any joint, muscle, or organ to function optimally, it requires the amount of space that it was designed to utilize. If dimension is diminished in length, width, or depth, there no longer is enough space for the body to operate more optimally. When body segments are compromised, a breakdown of tissue can result due to altered mechanics and undue wear and tear, as well as decreased efficiency of the musculoskeletal system and the internal organs. For this reason, we can see that observing space or dimension becomes just as important as observing alignment.

The length is the vertical distance; the depth, from the front all the way through to the back; and width, from one side all the way through the body to the other side. When we begin to look for dimensional relationships between segments, we describe them in terms of a part being longer or shorter, wider or narrower, or deeper or less deep, than another.

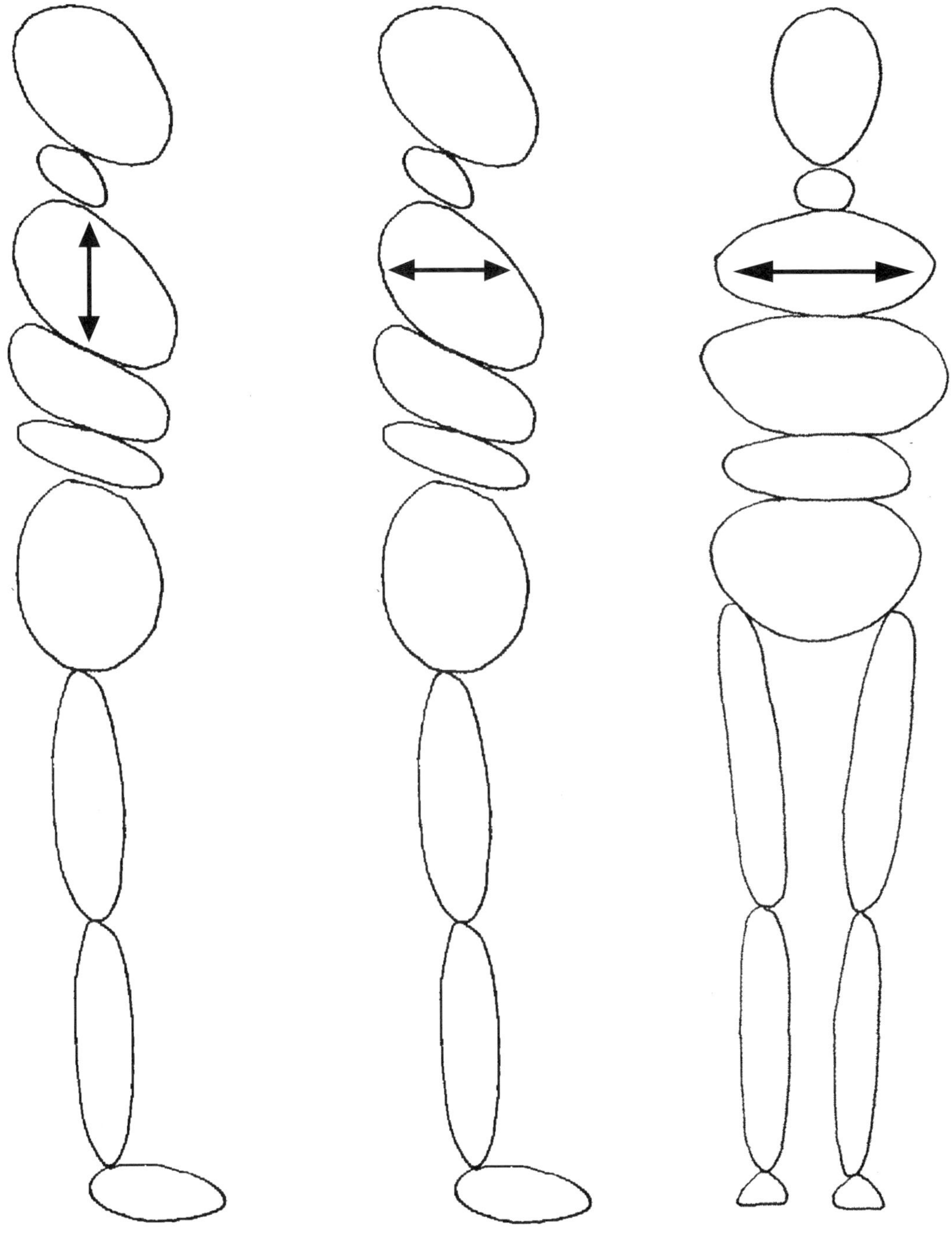

These combinations of the length, depth, and width of any segment give the body its shape or internal volume. The elasticity of soft tissues combined with the ability of the skeleton to change alignment enables each of us to move from possible compensations of constriction/restriction to a more optimal body shape and function. This gives each body segment three-dimensional integrity.

Take a look at these body patterns for the changes in dimension from the front-to-back views. Observe how the shape of one body segment reveals the necessary consequence in another segment(s).

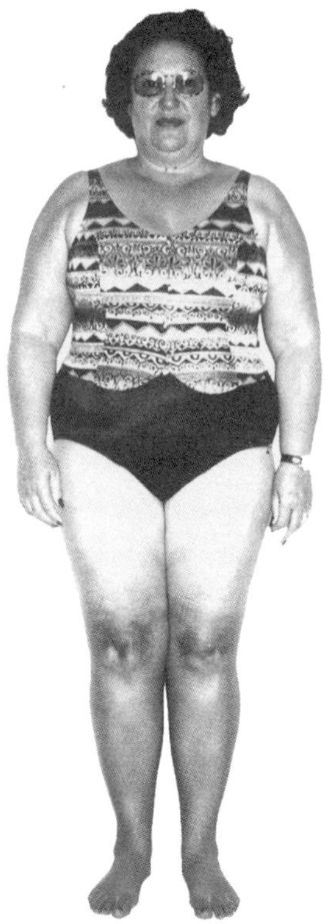

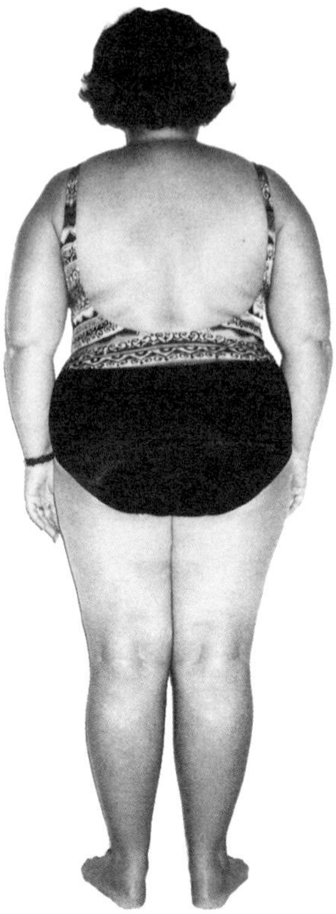

The natural shape and volume of the body segments are obviously different in the anterior-to-posterior-to-lateral views.

Neutral

1. Anterior view of the chest

2. Posterior view of the chest

3. Lateral view of the chest

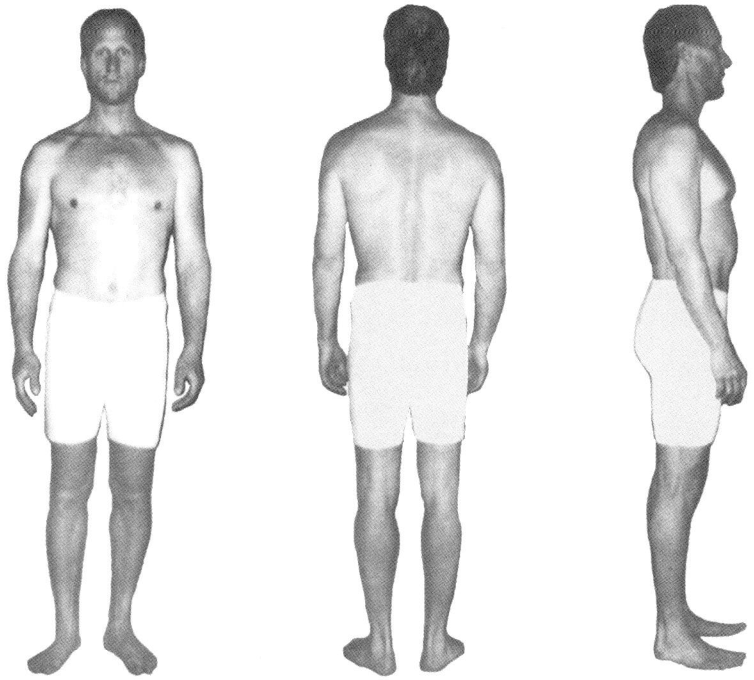

This pattern of the chest, with the person standing in his neutral position, displays a certain natural appearance and a matching of the chest's proportion from all angles.

Flexion

Many times, however, a quick glance at the frontal view of the client's upper chest reveals that it doesn't seem to match the size of the upper chest from the back view.

1. Anterior (narrow)

2. Posterior (wide)

3. Lateral (deeper posteriorly)

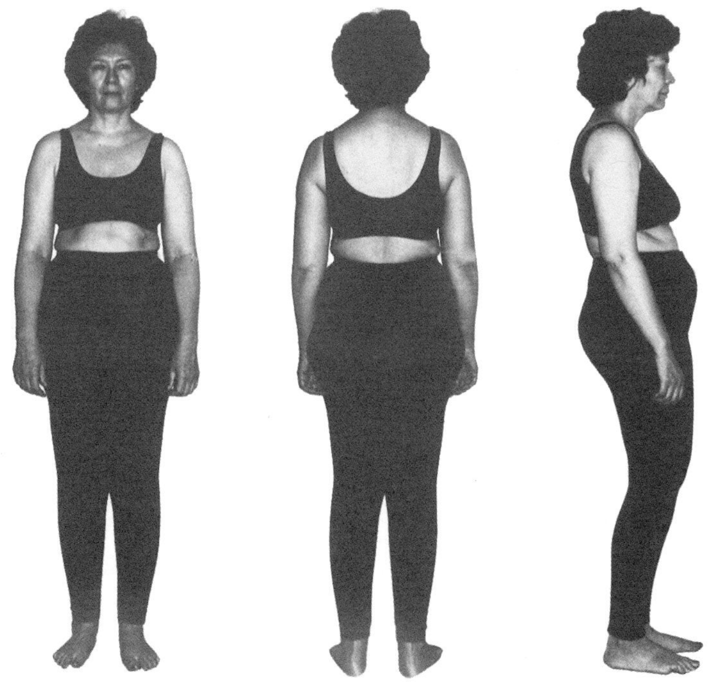

Extension

1. Anterior

2. Posterior

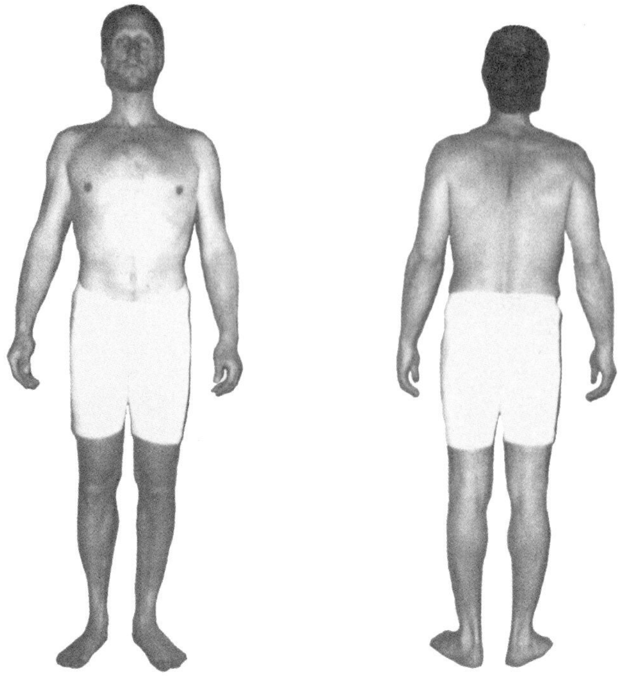

How does it happen that the front and back views of the chest can look like they belong to two different people?

The internal volume is shaped by the limited amount of three-dimensional space—length, depth, and width. For example, if an opera singer in training to increase lung capacity were to fill up his anterior chest by lifting it up and out, he would not have the same amount of volume in his posterior chest.

Notations for Dimension

The following notations can be used to describe the change in depth, width, and length:

Increased Note the direction of arrows.

Decreased Note the direction of arrows with the reversed brackets.

Examples

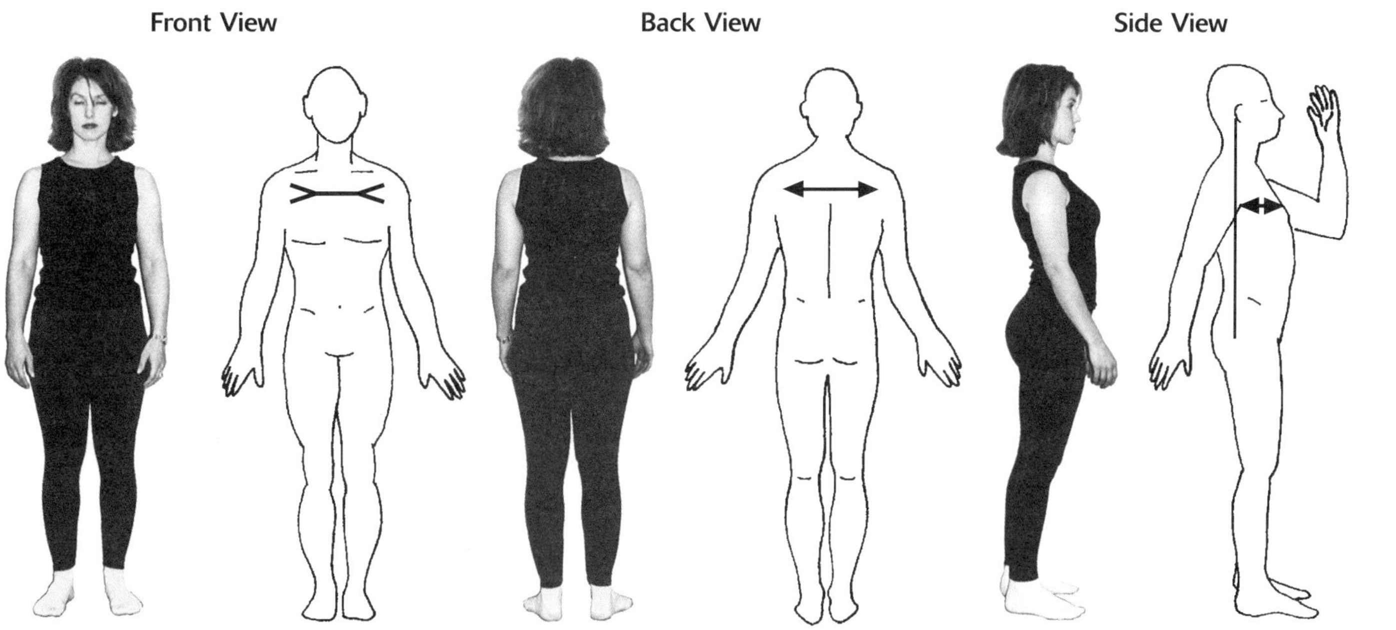

Demo

Seeing Dimensional Differences in Segments

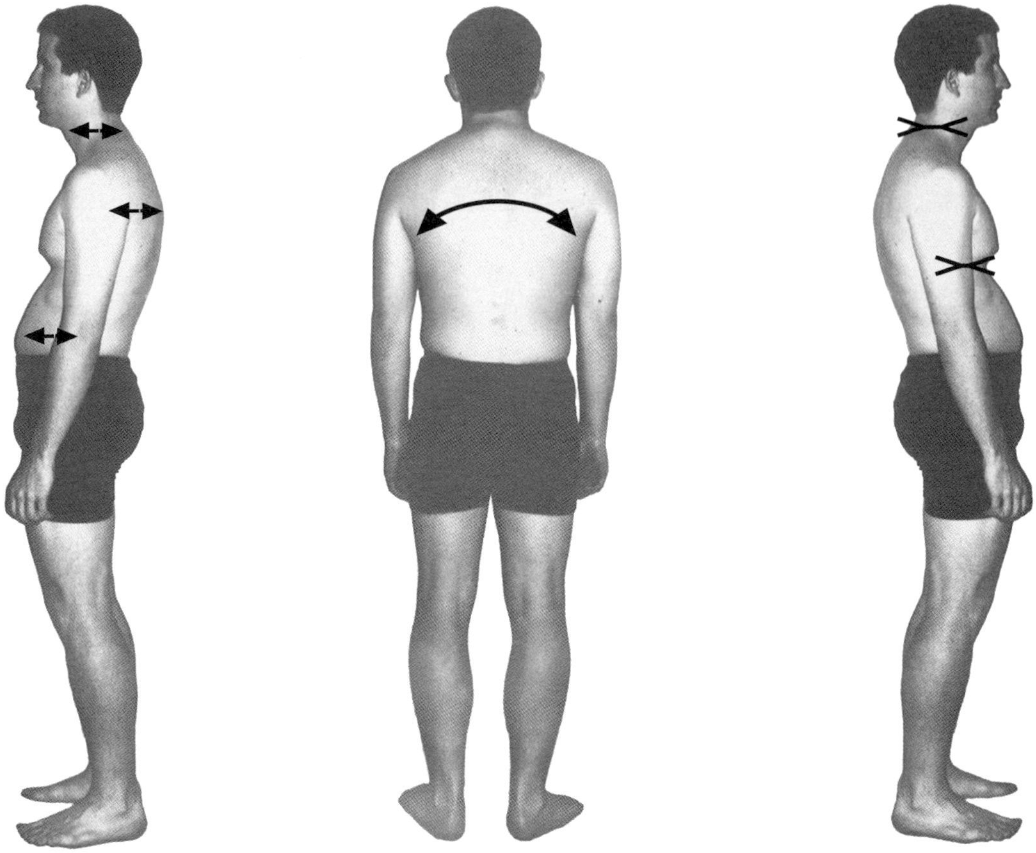

Exercise 3.1

Notation of Dimension

Notate the increased or decreased dimensions of the upper and lower chest.

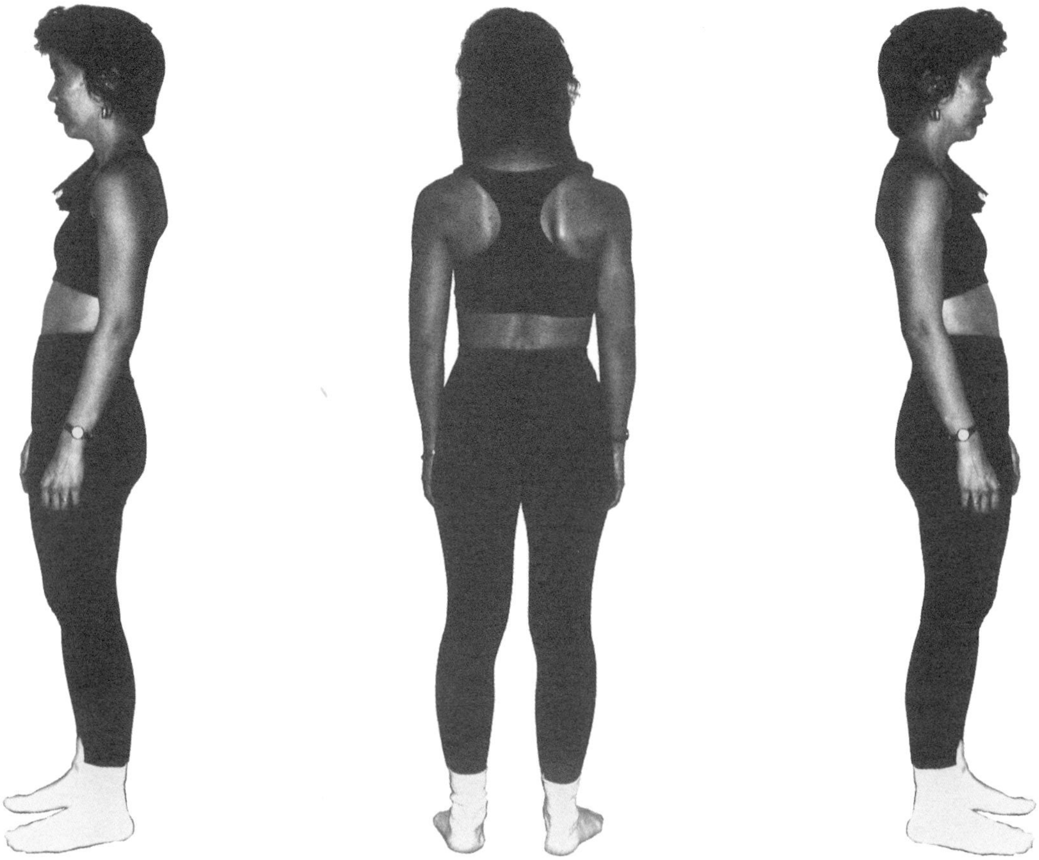

Exercise 3.2

Combination of Displacement and Dimension

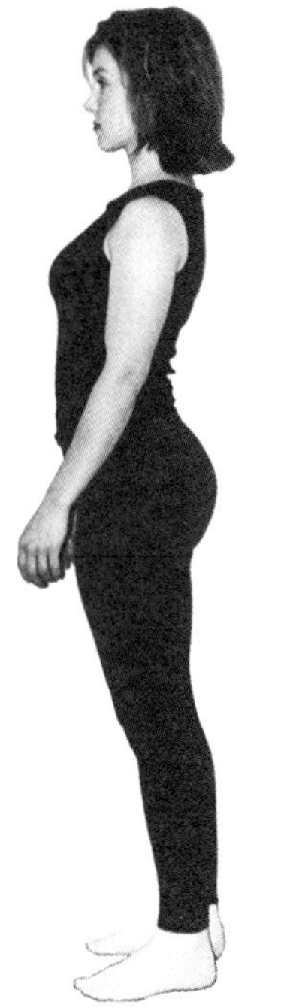

Placement
Dimension

Observe the tendency of the displaced segment to be in the same direction as the increased dimension.

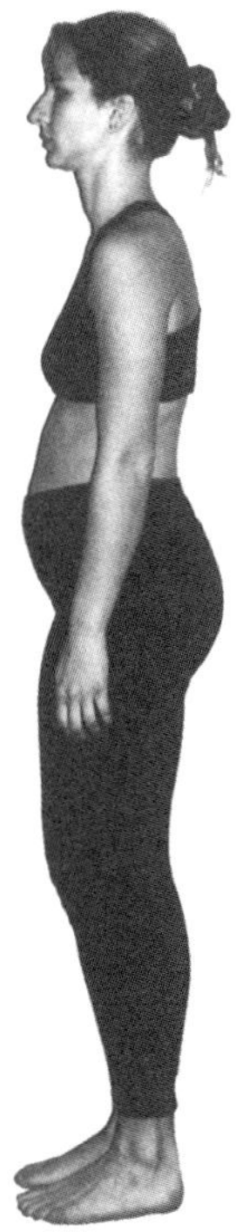

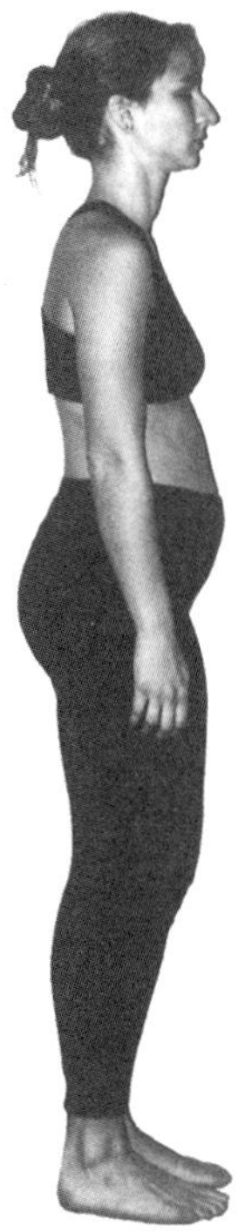

Notate the placement and dimension of the head, neck, upper chest, lower chest, and pelvis.

Flexion-Extension Patterns

Although there are many possible balances and counterbalances within the body, I began to notice two basic patterns: flexion and extension. There are 11 components that make up each pattern. When you can identify all 11 components, you will be able to quickly assess the client's body patterns.

It is useful to identify and learn each pattern in its generic form first, in order to be able to see the many possible and often subtle combinations of the two patterns throughout the body.

Flexion Pattern

The generic flexion pattern occurs when the whole body moves into a position that matches and possibly exaggerates the spinal flexion curves that occur naturally in the thoracic spine and sacrum.

Components

1. Body weight is centered posteriorly over the foot (heel).
2. Knees are flexed.
3. Upper and lower extremities are internally rotated.
4. Pelvis is tilted posteriorly.
5. Spine is flexed.
6. Breath is being exhaled.
7. Head and neck are flexed.
8. Eyes are faced downward.
9. Scapulae are protracted.
10. Body appears to be collapsing inward and downward into gravity.
11. Overall body length is decreased.

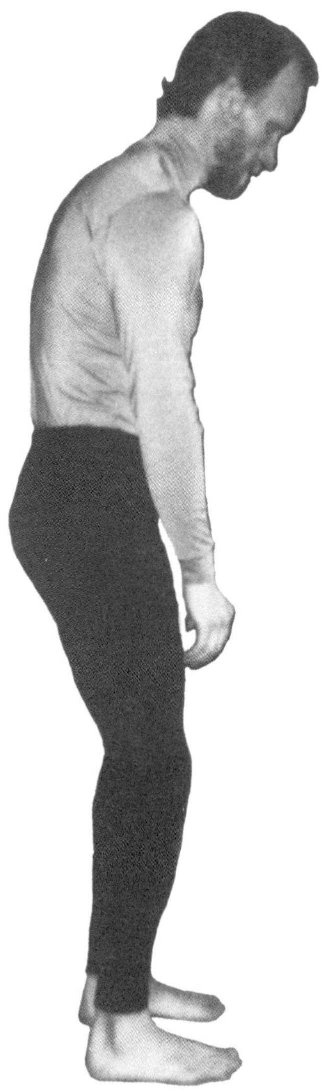

Extension Pattern

The generic extension pattern occurs when the whole body moves into a position that matches and/or exaggerates the spinal extension curves of the cervical and lumbar regions.

Components

1. Body weight mostly is centered over the forefoot.
2. Knees are hyperextended.
3. Upper and lower extremities are externally rotated.
4. Pelvis is tilted anteriorly.
5. Spine is extended.
6. Breath is being inhaled.
7. Head and neck are up.
8. Eyes are faced upward.
9. Scapulae are retracted.
10. Body appears to be using the push upward from the floor (GRF), or is holding against the pull of gravity.
11. Overall body length is increased.

Combinations of Flexion and Extension Patterns

The whole body is flexed. Head and neck are extended.

The whole body is extended. Head and neck are flexed.

Using the generic guideline of components, observe the following patterns. Which segments are in flexion and which are in extension?

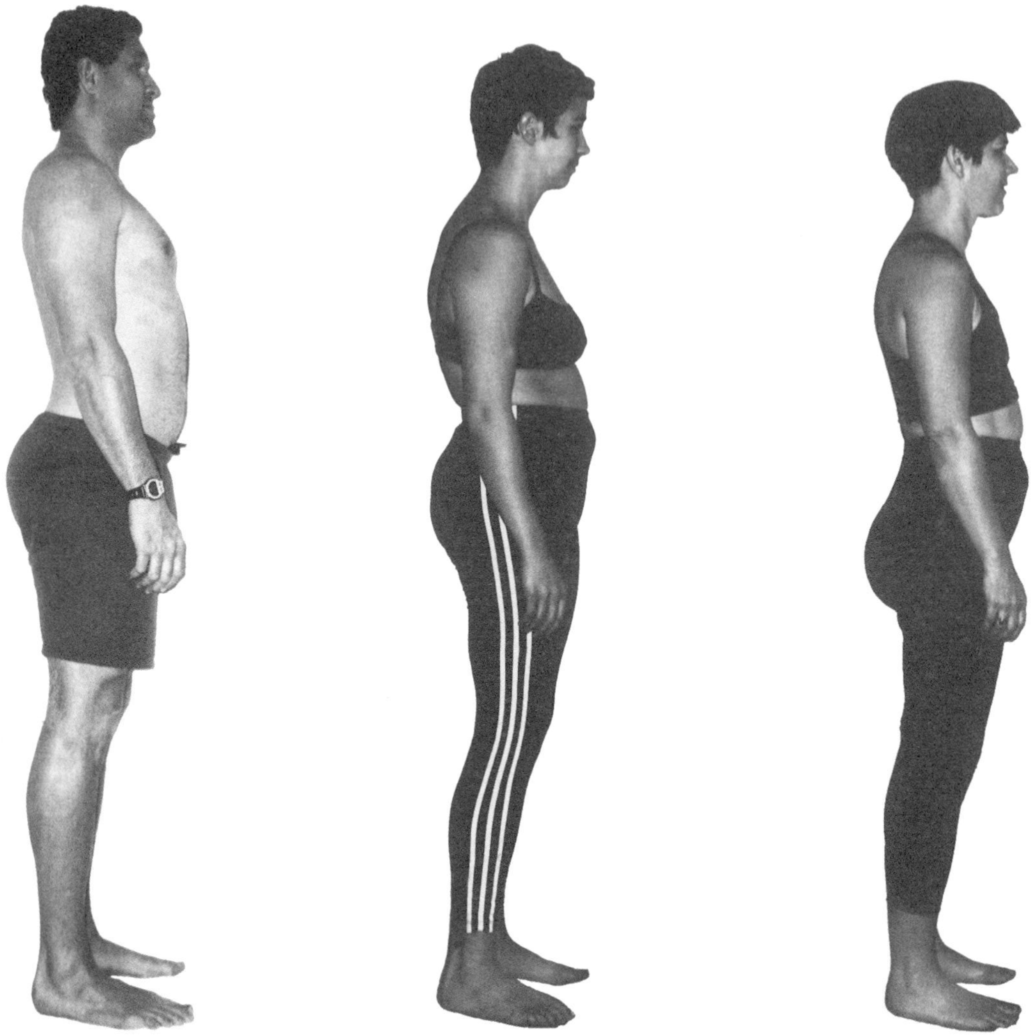

Relationships

1. Anterior/posterior relationships in alignment
2. Alignment and dimension
3. Alignment and apertures
4. Alignment and hypertone/hypotone
5. Alignment and hypermobility/hypomobility

Alignment Relationships

When we begin to see objectively what is occurring in the body spatially, we must learn to recognize the different relationships of structure to function, and vice versa. Our understanding of the effects of various factors on alignment, as well as the effects of alignment on these factors and of these factors on each other, will lead to a more accurate perception of the functioning individual as a whole system.

Although we will discuss these factors more in the next section, consider the scenario below that could cause them.

When a segment is displaced, the connecting joint and the segment above or below it can be affected in some of the following ways:

- The skeletal structure changes in position to accommodate the displacement.
- Indirect weight bearing occurs, which affects the direction of gravity and GRF.
- The joint and segment become compressed or stretched.
- Shearing occurs in the involved joint.
- Rotation can occur in the joint and soft tissues.
- The joint can become hypermobile or hypomobile.
- The soft tissue can become hypertoned or hypotoned.

Anterior-to-Posterior Relationships in Alignment

Compare the differences of the body's segmental relationship and weight bearing in the foot when we change the angle of the neutral from 0 degrees to 2–5 degrees of forward inclination.

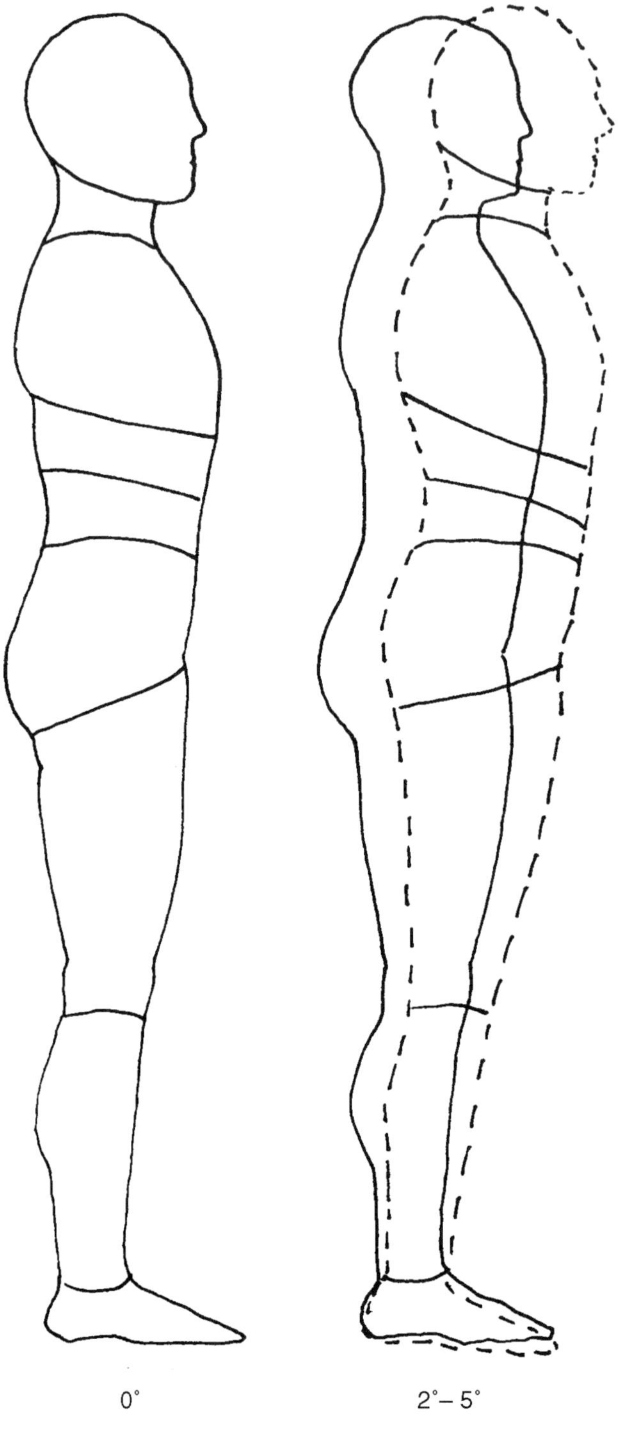

Body Landmark References

The following landmarks give us reference points to observe the angle of segmental relationships in the side view.

Because the body structures are inclined forward (as with the angle of the ribs), there will be a slight downward angle, posterior to anterior.

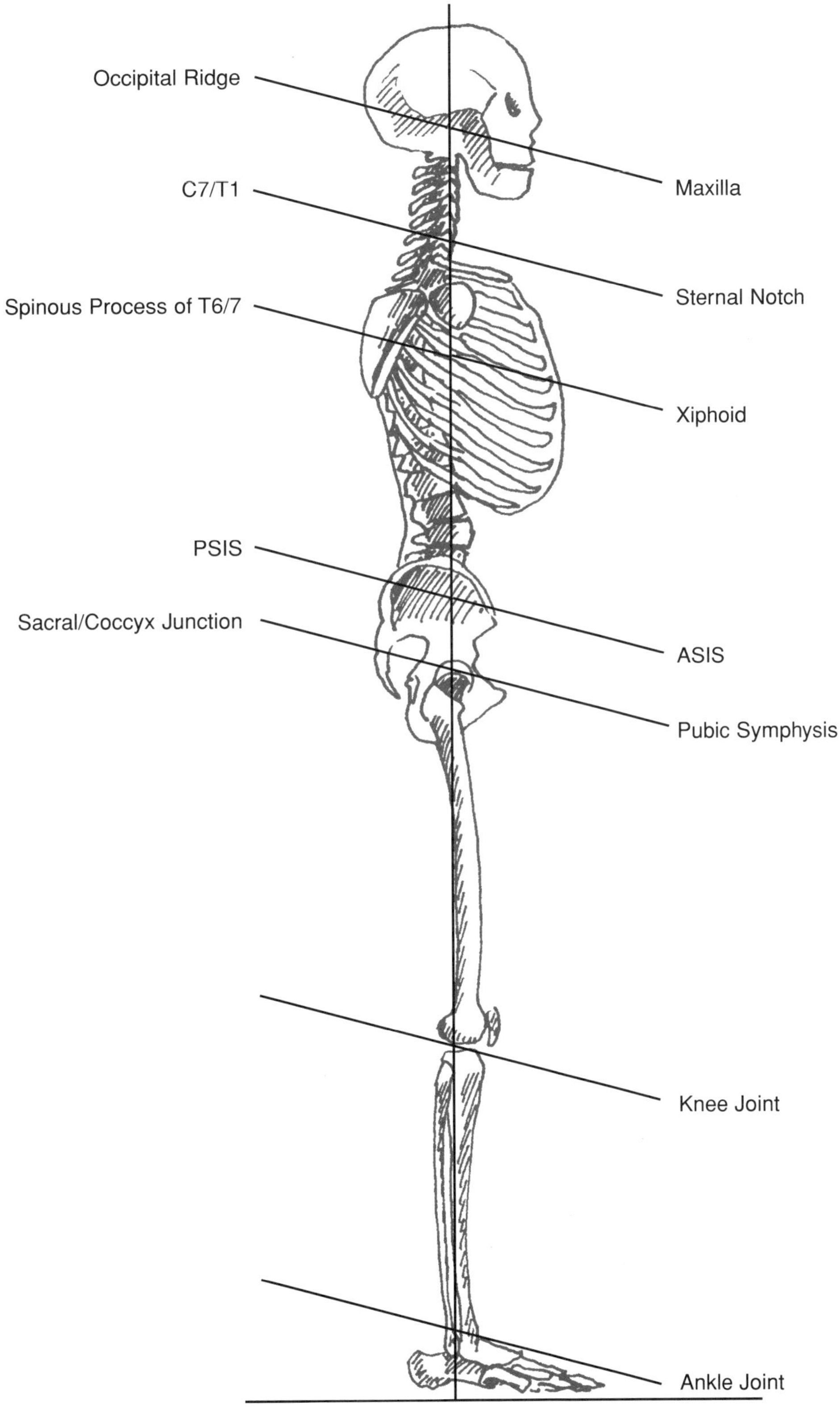

Using the previous guideline, mark the anterior and posterior landmarks on the following photos.

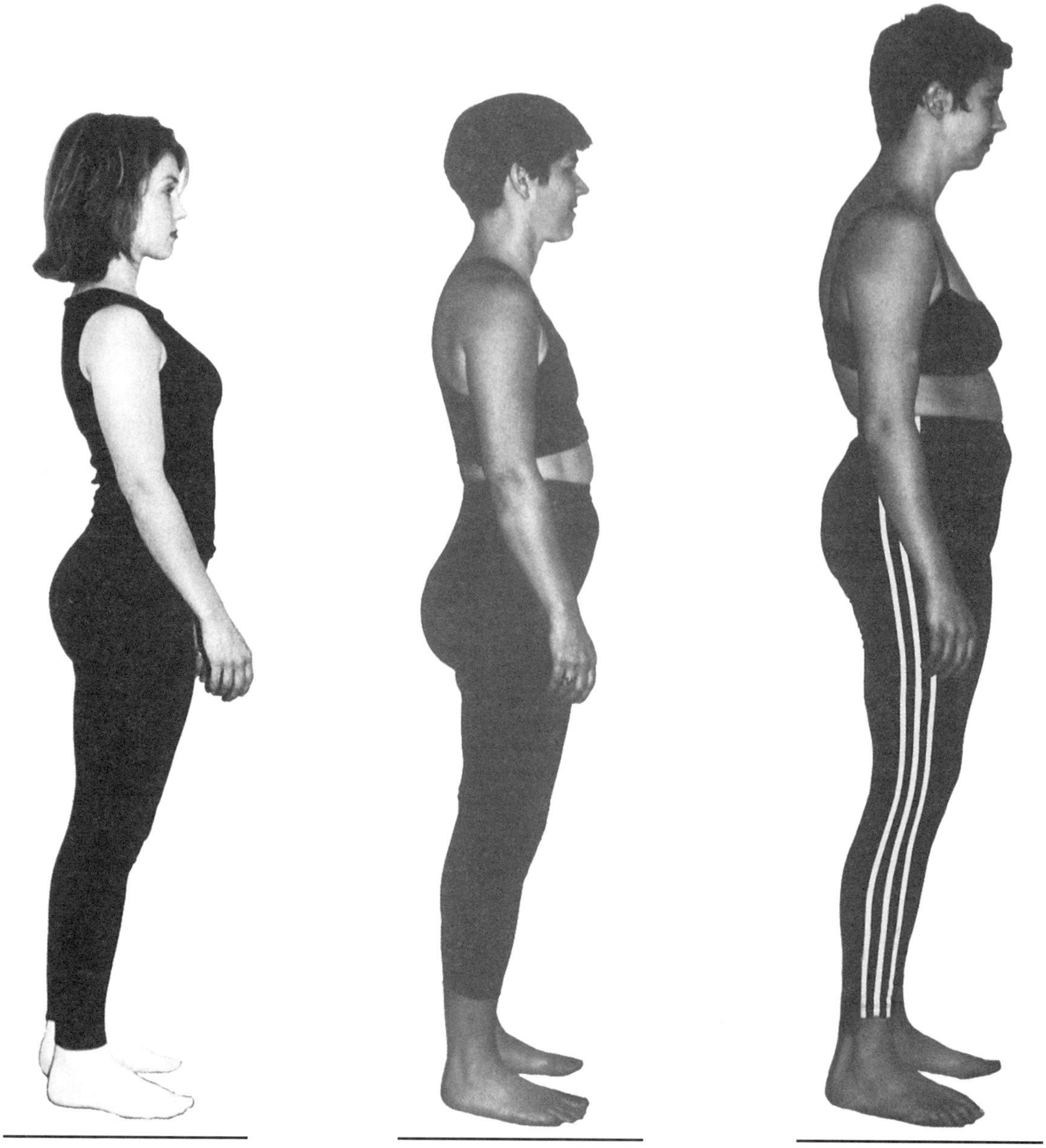

Draw a line to connect the landmarks.

What information does this give you?

How does this correlate with the placement of the segments anteriorly to posteriorly?

Example

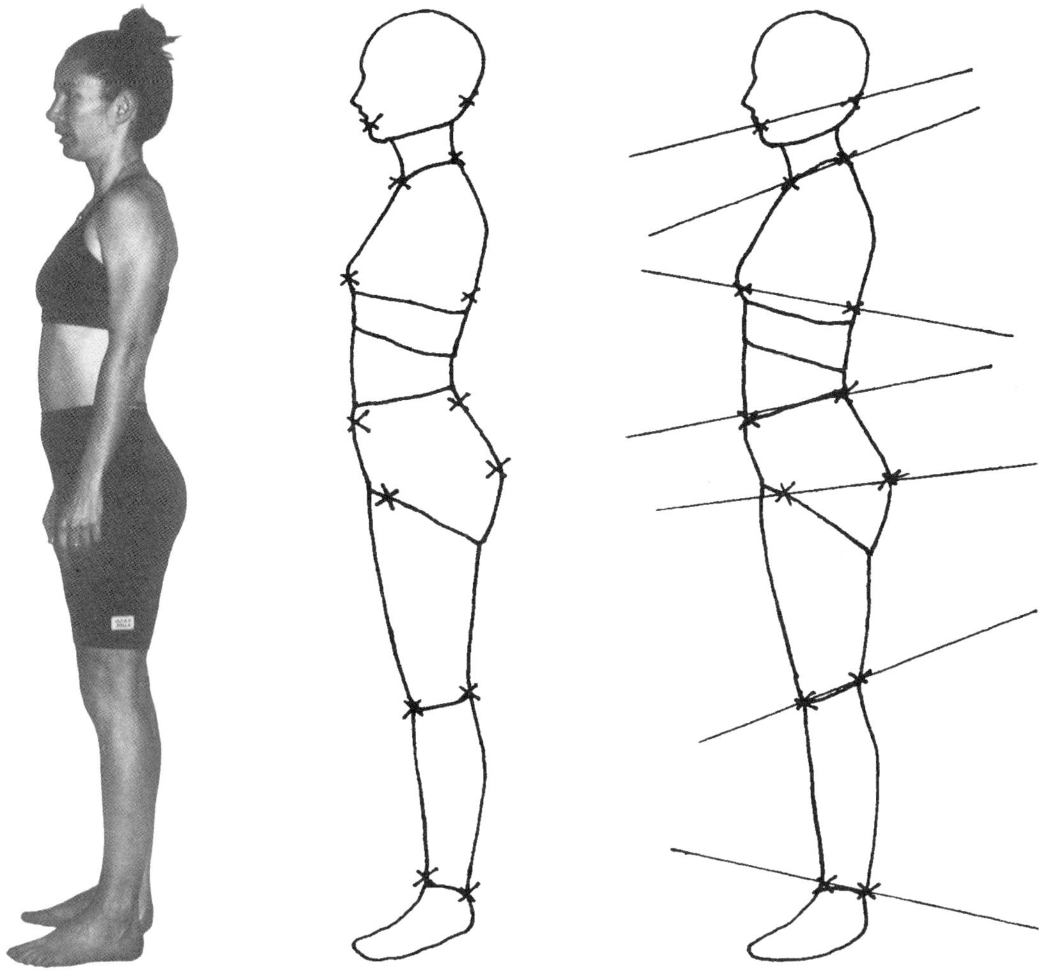

A. Photo   B. Segments outlined with landmarks   C. Landmarks connected

Can you identify the flexion and extension patterns?

The area where the segmental angle changes direction can be called a transition area. These areas are examples of compression, shear, or torsion.

Observe the combination of anterior-to-posterior placement with the up-to-down relationship at the joints.

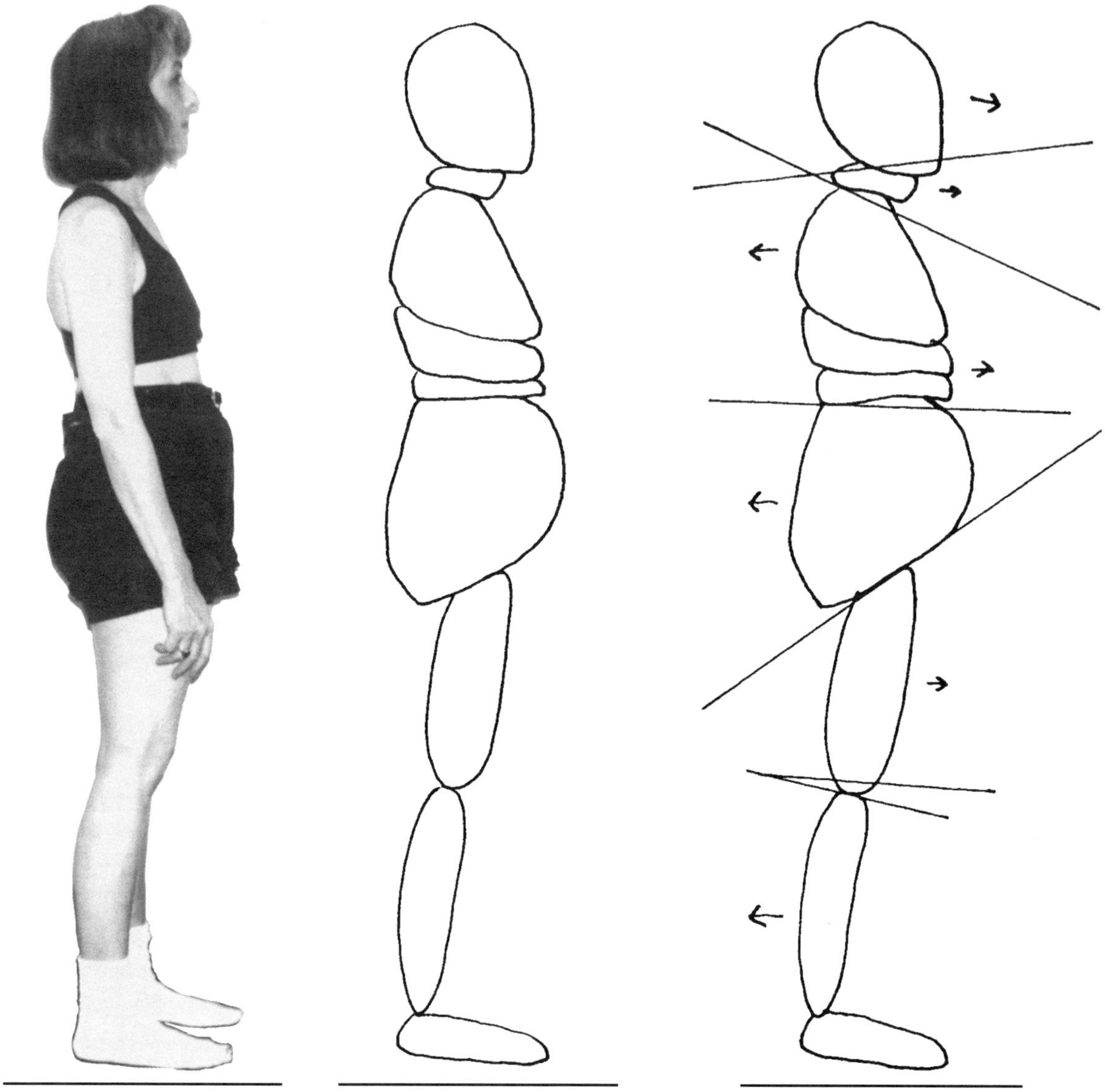

This combination of anterior-to-posterior placement and angles can provide immediate feedback on the primary areas of compromise. Each intersection of lines poses a strong possibility of compression.

Alignment to Dimension

There is a given amount of internal volume available to any body segment when the alignment of one segment changes. The length, depth, and width of that segment change shape to negotiate the alignment.

Generally, when a segment is shifted in one direction, the length, depth, and width increase in that same direction. For example, as one segment is shifted backward (posteriorly) the depth, length, and width of that segment are increased in that direction.

When one segment (the upper chest) is displaced posteriorly:

1. The depth is increased posteriorly.

   The depth is decreased anteriorly.
2. The length is increased posteriorly.

   The length is decreased anteriorly.

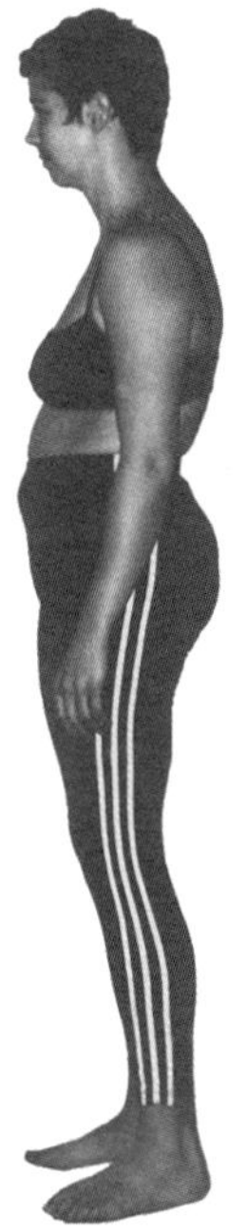

Looking at the front and back views, the lower chest is wider and longer in the back than in the front.

When one segment (the lower chest) is displaced anteriorly:

1. The depth is increased anteriorly.

   The depth is decreased posteriorly.
2. The length is increased anteriorly.

   The length is decreased posteriorly.
3. The width is increased anteriorly.

   The width is decreased posteriorly.

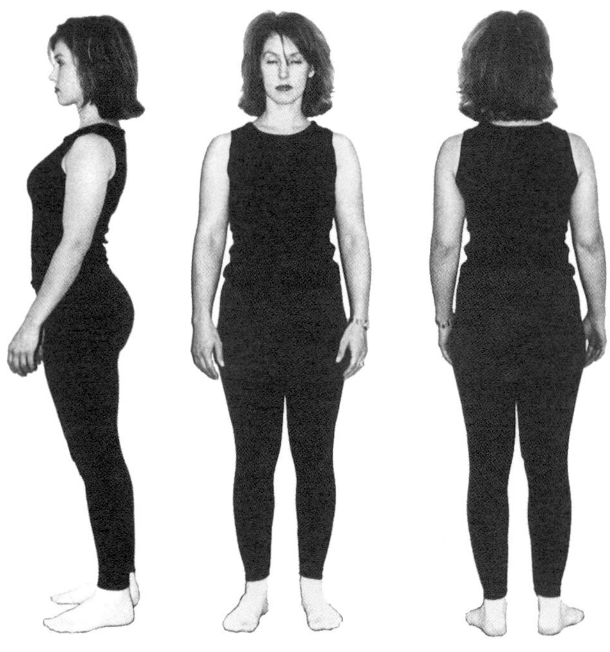

Demo

Alignment to Dimension

The upper chest is centered posteriorly to the neck and lower chest. Observe the relationship of dimension to anterior-posterior displacement.

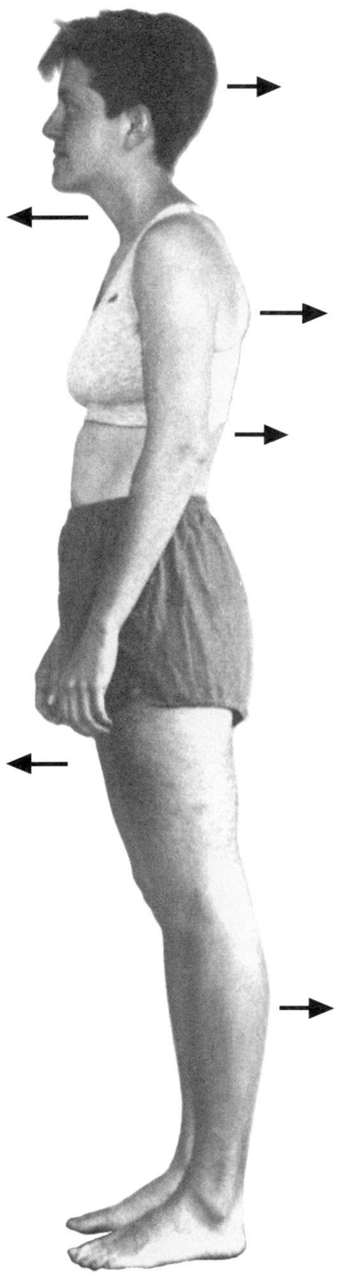

1. Increased depth of neck, anteriorly.
2. Increased depth of upper chest, posteriorly.
3. Decreased depth of pelvis, posteriorly.
4. Increased depth of upper leg, anteriorly.
5. Increased depth of lower leg, posteriorly.

Exercise 3.3

Observe the tendency that as one segment is displaced, this increases the depth, length, and width in the same direction.

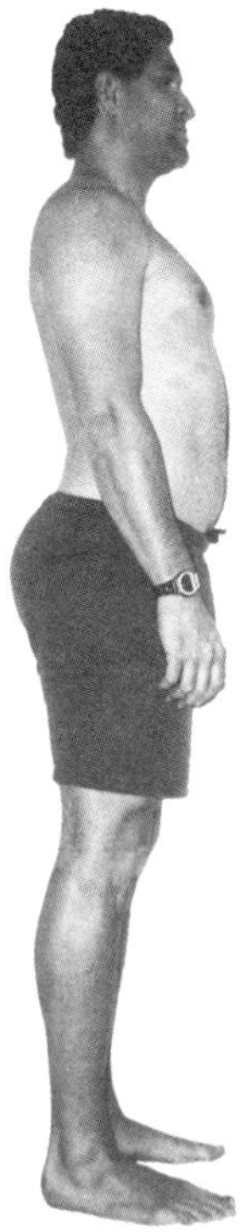
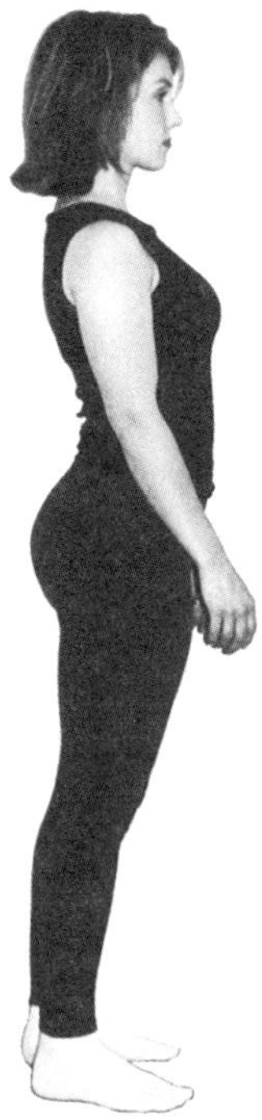
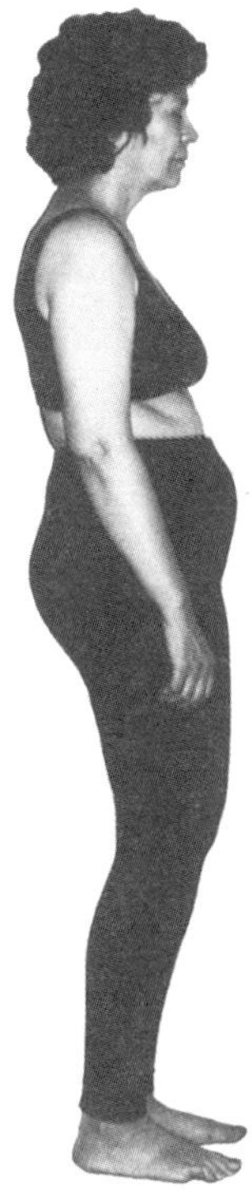

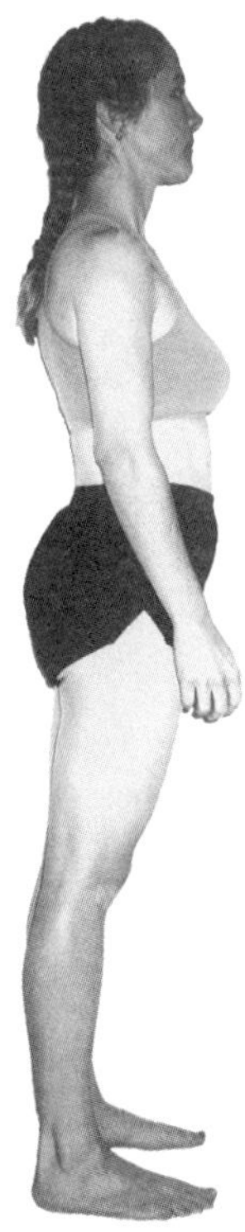
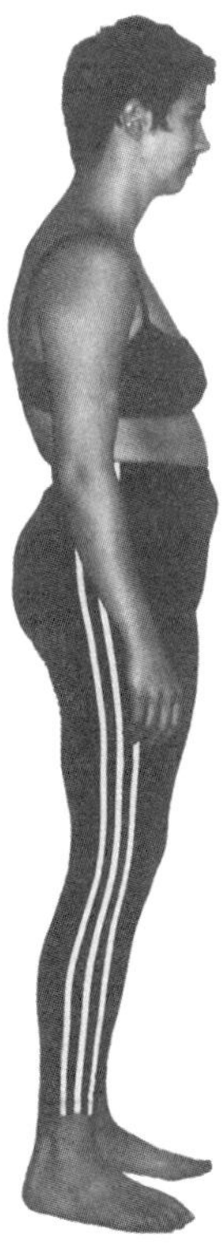
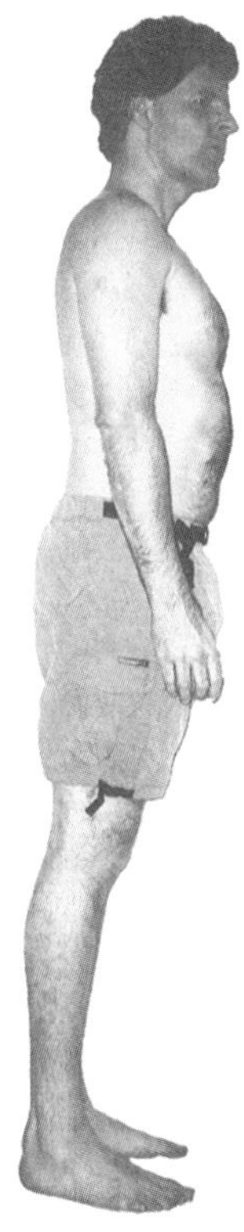

Factoring dimension into one's plan can affect the sequence of treatment and exercise.

Aperture—Alignment and Aperture

Aperture is a term that Aston-Mechanics uses to refer to the cross-sectional areas of the body. An example of an aperture would be any two-dimensional cross-section of a segment. These intersections of segments behave like the iris of the eye or the aperture of a camera. An aperture can increase, decrease, or change shape depending on its relationship to the adjacent segments. The aperture can be observed in any of the three planes, in order to indicate the two-dimensional volume of any segment. For now, consider the aperture through the transverse plane of the body segments.

When the body is in its neutral alignment, the apertures are at their optimal functioning capacity for direct weight bearing among the segments. Gravitational and ground reaction forces of the body find their pathway through its combined apertures.

Cross-section of the leg

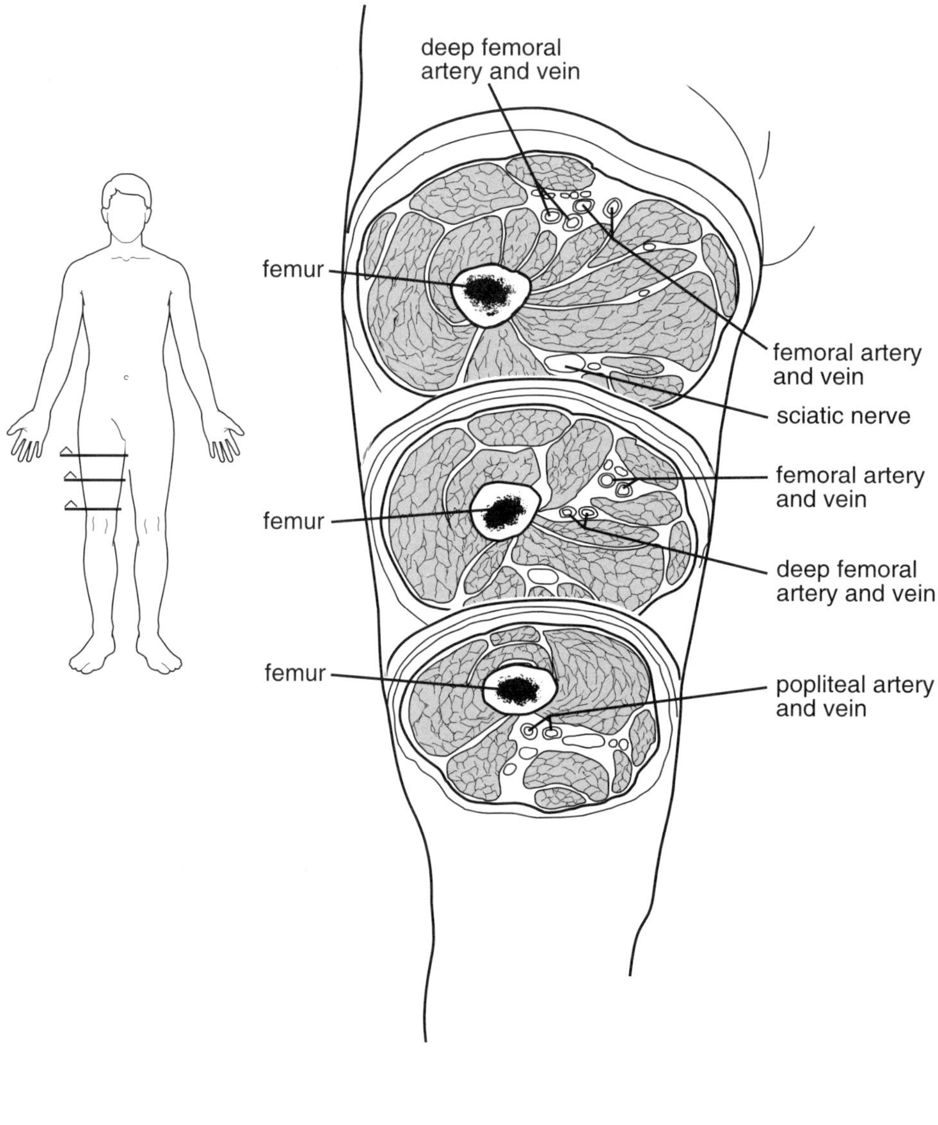

Many times the rehabilitation process is prolonged when patients do not know how to match their available apertures for weight bearing to their unique physical limitations. Instead, they prematurely force their bodies into ideal alignment.

When a segment is displaced, the apertures of the segmental intersections will be reduced or expanded according to the amount and direction of the displacement.

Two examples of this are in overhead views of the body that follow. The illustration on the left shows a more neutral position while the one on the right demonstrates a body with anterior-to-posterior displacement.

From overhead, notice that displacement anterior to posterior, right to left, reduces the surface area or BOS for direct weight bearing.

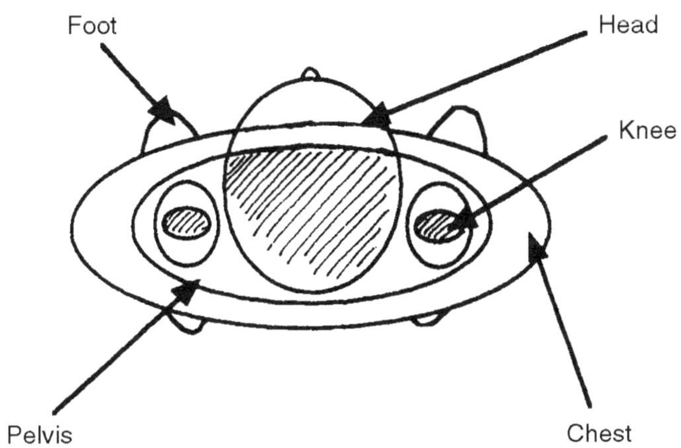

Neutral

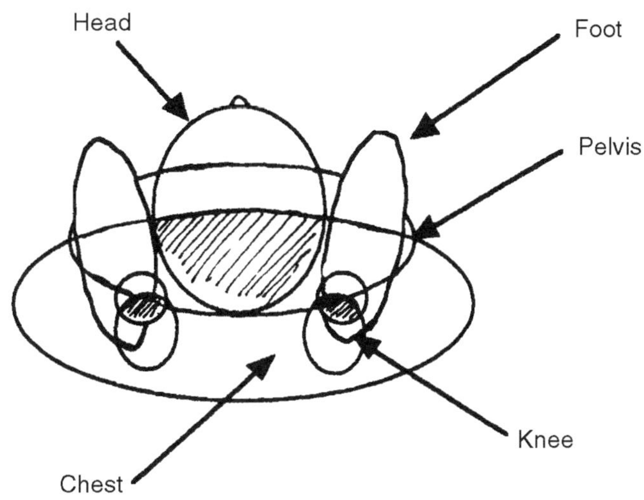

Displacement of segments

The shape of the aperture will then be affected accordingly. Compression often will increase the diameter of the aperture, and distraction or tensile force will decrease the size of the aperture.

Now let's look more specifically at how displacement from one segment to another affects the dimension and function of a specific aperture. Observe the cross-section of the leg in the drawing below. We can assume that this is an example of someone standing in the neutral position. Imagine how this cross-section would look if the person were standing in a flexion pattern with weight bearing over the right foot. This would most likely increase the anterior dimension of the right thigh, and the same components of muscle, bone, nerves, and all soft tissues likely would be changed in position and dimension by the biomechanical stresses.

Cross-section of the leg

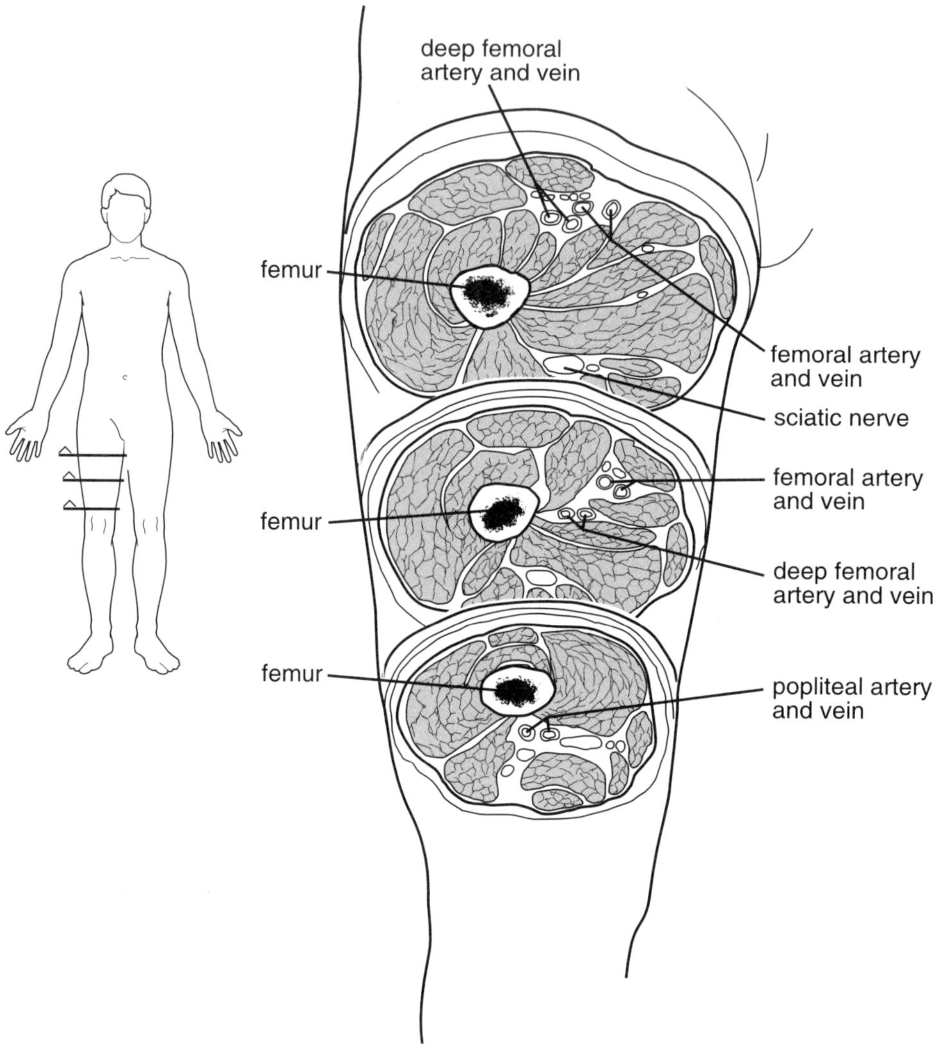

Observe the following photos for displacement and change in dimension, as well as for possible consequences such as shear, compression, and torsion.

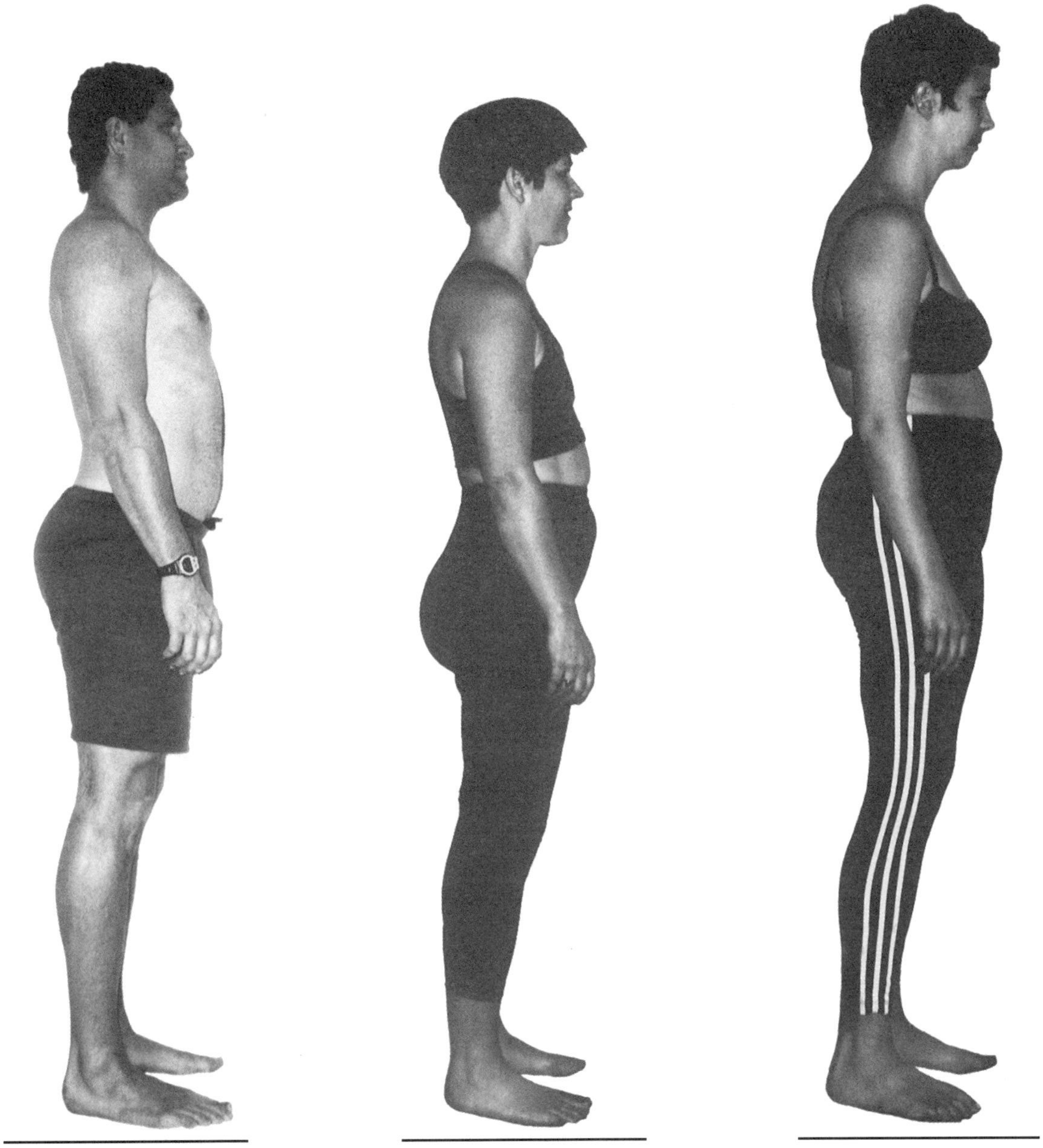

Relationship of Alignment to Tone

Tone

The tone of the muscles is the relative presence or absence of contraction, tension, resilience, or holding within the tissue.

It is the appropriate tone for any given area and the whole body, with respect to the history of injury, age, activity level, health, and so forth. Senior citizens are finding more value in weight training these days, but one would not expect the muscle tone to be the same for a 95-year-old and an 18-year-old. There are exceptions, of course.

Proportional tone, which is unique to each individual, is the amount of tone necessary to stabilize the skeleton and allow ROM. Also, it is the amount of tone appropriate for optimal function for each person.

TONE CONTINUUM

1	2	3	4	5	6	7	8	9	10	11	12
Hypotone					Proportional Tone						Hypertone

Hypotone is a laxity of tone, or the presence of less than proportional tone. Hypotonic tissues are slack, overly responsive to forces moving the body, and are lacking in stability.

Hypertone is excessive tone in tissues, or more than what is proportional. Hypertonic tissues are overly contracted and dehydrated, lacking the resilience to respond to the forces of motion, and to expand, stretch, or otherwise change shape.

When a joint is surrounded by hypotonic tissue, the joint lacks stability and is at risk for easy displacement. When a joint is surrounded by hypertonic tissue, the joint can become compressed, and normal joint mechanics are altered. The combination of the continuous effort of the hypertone and the instability from the hypotone can cause and reinforce the positions of displacement.

Being able to assess the amount of hypertone and hypotone in the body by merely observing it can pose a challenge. Adding palpation skills can be useful in assessing the tone accurately to confirm your observations.

Previously, we observed that:

1. When body segments are displaced in one direction, they are balanced or counterbalanced by other segments being displaced in the opposite direction.
2. This balance/counterbalance is created by increasing dimension in the displaced segment(s) in the same direction as the displacement and by decreasing dimension of the segment(s) in the opposite direction.
3. The balance-to-displacement relationship is influenced by an increase in tone. In the following photos, the clients are leaning quite forward (anterior), with little displacement posteriorly. How can they do this? Generally, they can achieve balance in this position by increasing tone posteriorly (hypertonicity) as well as sometimes gripping the toes.

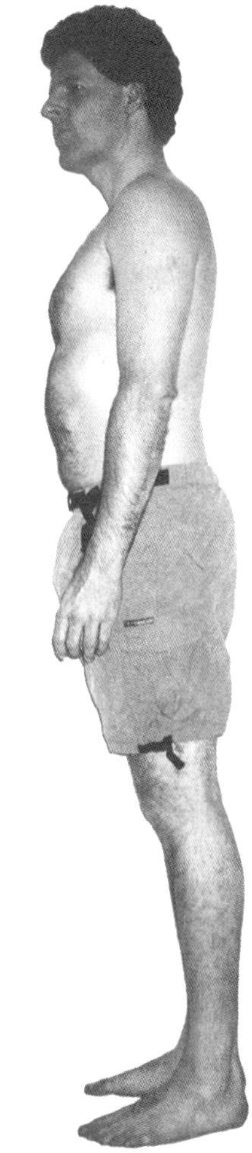

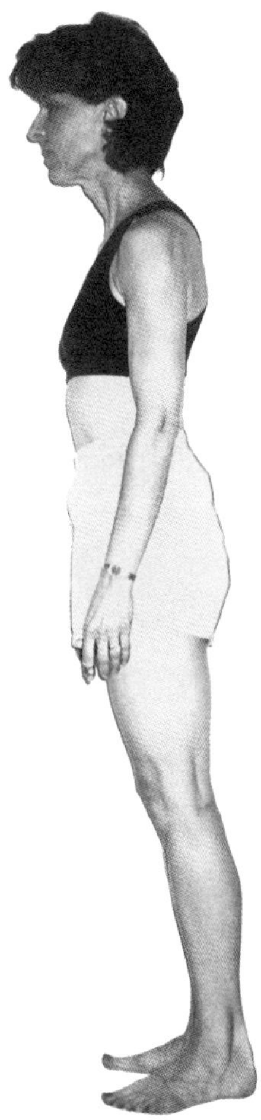

Many times, a joint is compressed because its length has been decreased by hypotonicity (not enough tone to stabilize the joint) in the surrounding area. When you add exercise and movement education to teach the client an improved neutral position that uses the alternating forces of gravity and GRF, the result can be an increase in the length of that segment, which, in turn, increases the tone in the surrounding tissues. This result creates a better point of departure for deciding and prioritizing what strength exercises or manual therapy techniques would be best to follow. In this case, the effect of doing strengthening exercises around a compressed joint may cause additional stress or compression.

Relationship of Alignment to Mobility

Mobility

Mobility is the amount of movement, or ROM, occurring in a particular joint. It can be observed on a continuum of movement, in relation to existing conditions of the body.

MOBILITY CONTINUUM

1	2	3	4	5	6	7	8	9	10	11	12

Hypomobility — Proportional Mobility — Hypermobility

Proportional mobility is the amount of movement available at a joint that is consistent with the joint's normal function when considering such factors as the individual's age, health, and condition.

Hypomobility is limited ROM and the inability to move in response to forces.

Hypermobility is increased ROM, which can produce instability. When a joint is hypermobile, it can be easily displaced and, therefore, is at risk of injury.

Example

A. A more neutral position creates more even tone and mobility through the ankle, knee, and hip.

B. The flexed knee position increases tone and can decrease mobility through the joints.

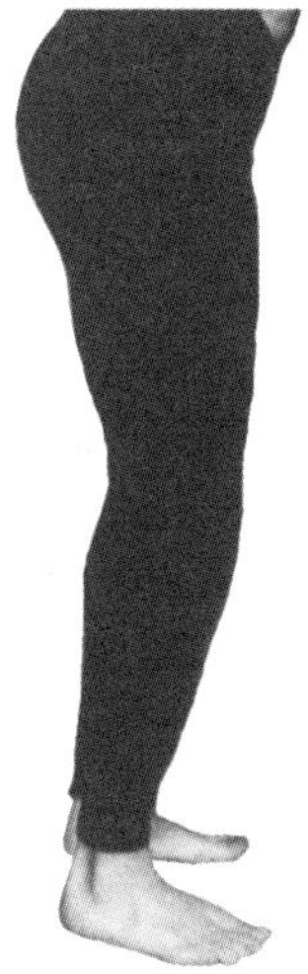

C. The hyperextended knee position creates uneven tone and mobility.

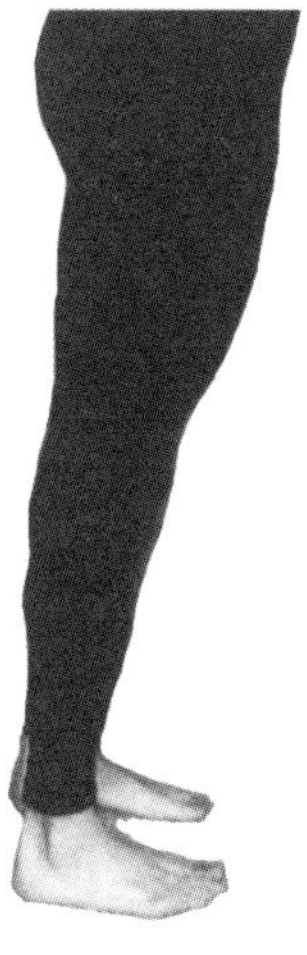

Let us observe the hypermobility/hypomobility of the hip joint as it relates to the pull of gravity.

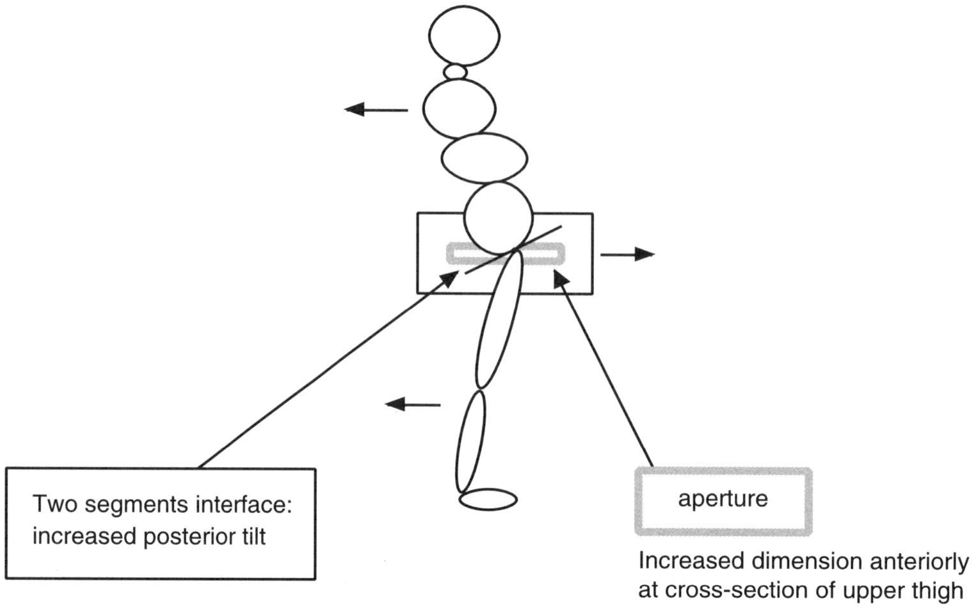

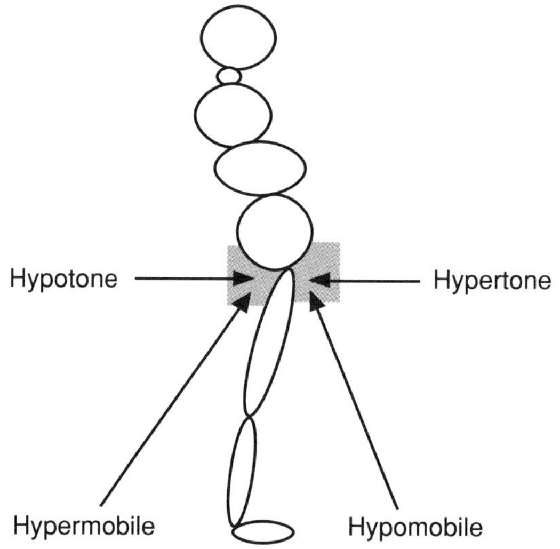

In this postural pattern, the gravitational force would move the pelvis forward and push the head of the femur at a sharp angle into its end of range. The result of this position is an overly stabilized hip joint, anteriorly. The gravitational forces would continue to pull the weight of the chest downward and, because of the lack of tone of the posterior hip muscles, continue to push the femur forward, which results in hip-joint hypomobility, anteriorly, and a hypermobility, posteriorly. (If you were to test the joint unto itself in non-weight bearing, you might find the opposite pattern, as compared with our example of testing for functional mobility.)

Compensations

Compensations are adjustments the body makes in order to accommodate a lack of balance. Compensations often happen as a subconscious attempt to match a limitation. They may increase or decrease displacement or cause one area to overwork while another area does not work hard enough.

Because the body is mobile, balance is constantly in flux. Balance is influenced by habitual ways of moving, repeated actions or activity, and by specific injuries or events. The inability to bear weight because of a broken toe or ankle sprain, for example, will cause the body to adjust or develop compensations in order to maintain balance.

Sometimes the source of an impact trauma creates an emotional response (e.g., loss, fear, anger). This is an example of when the original cause, emotional or physical, is the action that creates a reaction in the body, which, in turn, creates the need for compensations.

People have many options for negotiating the recording of physical and emotional history that may have caused various misalignments. One possible option is for the muscles to increase their effort to hold segments in balance with increased tone. Another option is that adhesions are formed in the myofascial structures to help support or reinforce segments out of alignment. Both of these options lead to uneven tension around joints that, in turn, leads to uneven patterns of movement and strain. Often, this cannot be seen in the photos but can be palpated. That is, the therapist can feel myofascial restrictions, increased tone, and joint hypermobility and hypomobility.

Another option available to restore the balance of segments occurs in the displacement of parts. As one segment is placed forward, another shifts backward, and balance is achieved. Counteralignment helps achieve balance.

For example, when you see a lateral shift of the pelvis in one direction, you might see a lateral shift of the shoulders/upper chest in the opposite direction. This is counterbalance, and it occurs in order to maintain balance of the body's segments around its center.

There also are many combinations of segmental displacement counterbalanced by increased tone in an opposite area.

Symmetry often is linked with the term *balance*. Both can imply an arrangement of parts and details to be equal on either side of a median line—to be a mirror image of each other.

Balance also can mean that dissimilar or opposing forces offset each other.

It seems obvious that most bodies are in balance, meaning that our body compensations keep us standing. However, this may not produce the best available balance for the short or long term. I define balance as the moment-to-moment negotiation of asymmetrical differences for the best function available.

Some of the consequences of a more balanced posture (unique to each person) are:

1. The placement of body segments allows for optimal ROM at the joints.
2. The weight bearing is more direct, which can positively utilize the compression and distractive forces.
3. The effort for work (muscle use) is more evenly distributed throughout the body.
4. Because the three dimensions are more optimal, the internal volume is more optimal, which improves the functions of breath, digestion, circulation, etc.
5. The muscle tone is strong enough to support as well as stabilize structure, and resilient enough to allow full ROM.

As you increase your observation skills, this allows you to recognize when each person finds their unique balanced posture, which changes from moment to moment.

As the body approaches its optimal posture, the position, space, effort, movement, and physiology are more optimal. The consequences of these relationships will determine the body's ability to feel increased awareness, support, stability, resiliency, mobility, and the possibility of a full range of expression of thoughts and emotions

Sometimes the body patterns create shapes that may or may not be accurate patterns of expression. For example, people experiencing grief may take on a slouched posture. Perhaps they took on the pattern many years ago and it has remained with them. Sometimes, people will make an interpretation of someone's patterns and relate to them accordingly. People are unique in their complexity. Even if people have had similar experiences, the how, the when, and the content of those experiences combine to make the results quite different.

Exercise 3.4

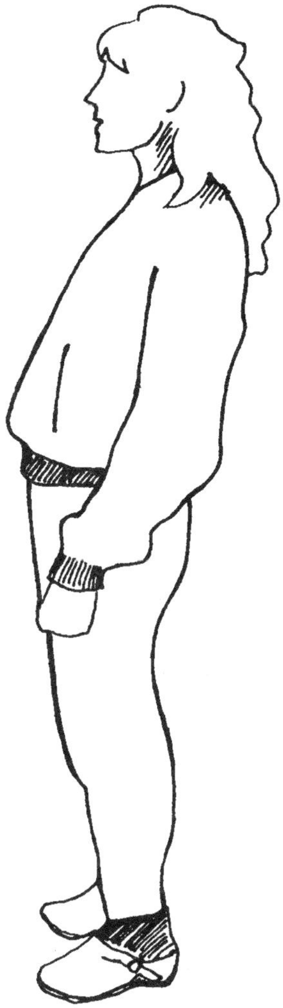

Observe the abstract body drawing below and answer the following questions.

1. What gets your attention first?
   a. The alignment
   b. The dimension
   c. The expression
2. What segment gets your attention the most?
3. That segment (from question No. 2) is in relationship to what other segment?
4. What primary track does this body pattern create?
   a. Extension/flexion pattern
   b. Rotational pattern
   c. Pattern of displacement—anterior to posterior, right to left and up to down
   d. Dimensional changes
5. When you have identified your primary track, what seems to be the major consequences in:
   a. The biomechanical forces of:

      Compression

      Tensile

      Shear

      Torsion
   b. The increase and decrease of hypertone and hypotone
   c. The path of direct weight bearing
   d. The path of translation of gravity and GRF
   e. Possible challenges this person may experience
   f. Compare the above information with the client's complaint that you should have recorded at the beginning of the session.

Possible Assessment

1. The alignment gets my attention.
2. The posterior alignment of the upper chest is the segment that gets my attention the most.
3. The posterior alignment of the upper chest is in relationship to the hip joint placed anterior to the ankle.
4. This body pattern creates an increase in the length of the anterior chest, in relationship to a decreased length of the low back.
5. Possible challenges:
    a. Compression or impact to the low back
    b. Decreased breathing capacity in the posterior chest
    c. Decreased support for weight of the head and, therefore, increased tension on posterior neck, upper back, and shoulders
    d. Shear through the viscera and the diaphragm
6. Compare your assessment to the client's complaint or focus:
    a. Low back pain at work
    b. Feeling stressed or frustrated

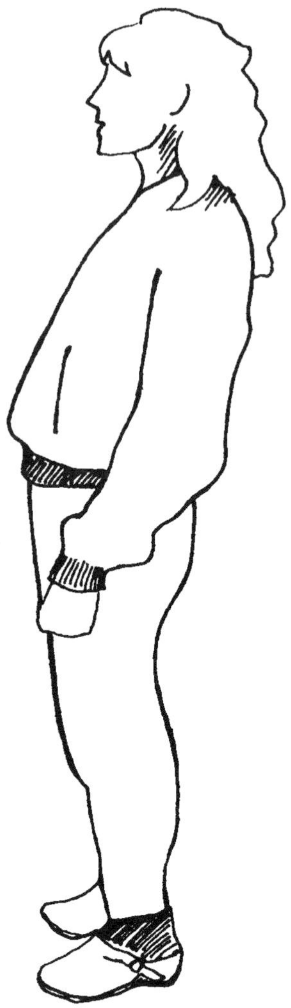

In relation to this illustration, there are two patterns that sometimes are overlooked.

1. The combined anterior placement of the hip and posterior-rotated pelvis move the ischial tuberosities close to the femur, which decreases the posterior depth of the hip joint and, therefore, decreases support for the posterior depth of the segments above.
2. Sometimes, therapists assess a client as having an increased lordosis when, actually, it appears more lordotic because of extreme posterior displacement of the lower and upper chest combined with anterior placement of the hip joint.

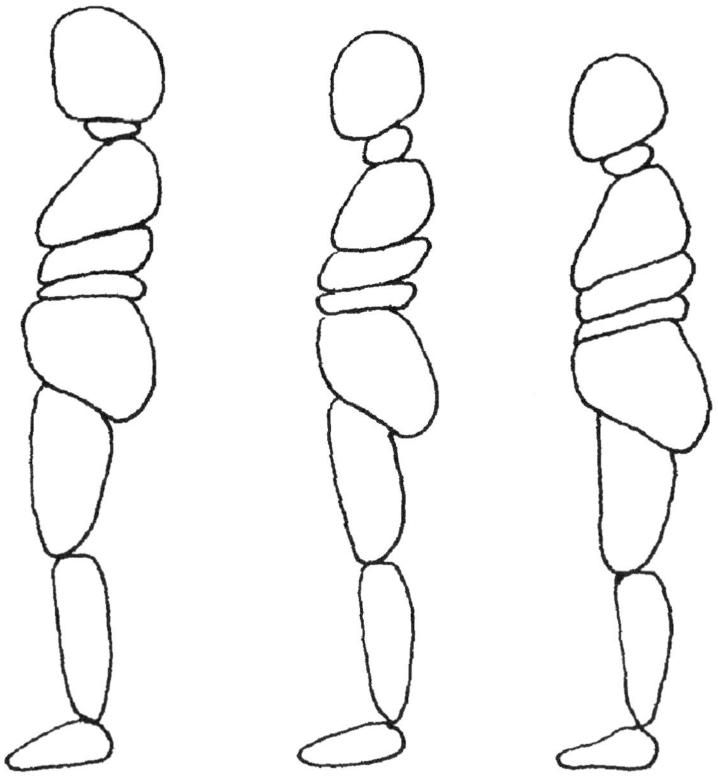

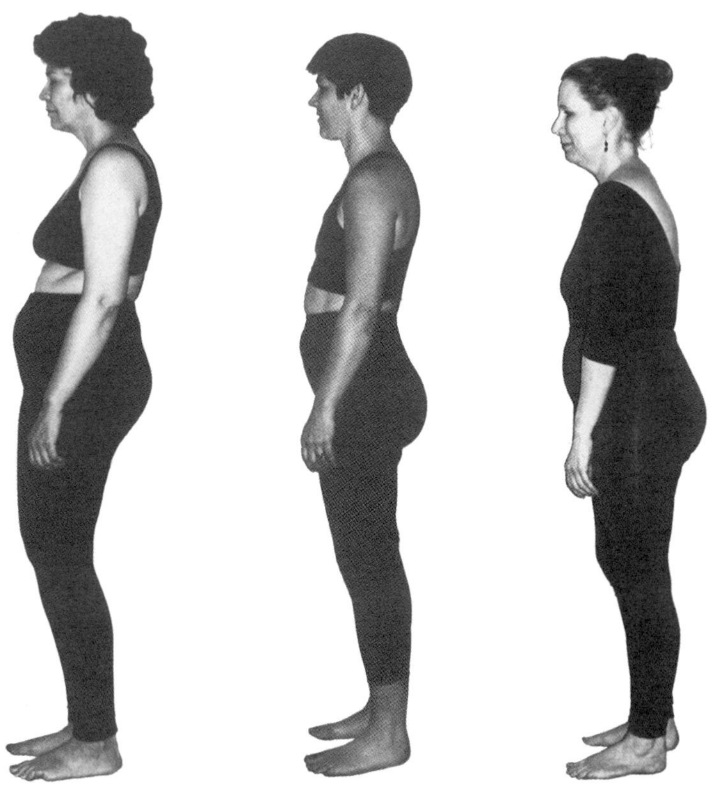

Perhaps by now you are able to assess as much information from the photo as the abstract drawing. It may be easier to see segmental relationships with the abstraction. Therefore, you may find it quite helpful to continue outlining the body segments of your photos onto a transparency.

Exercise 3.5

Observe the whole body patterns and consider possible consequences to the segmental relationships.

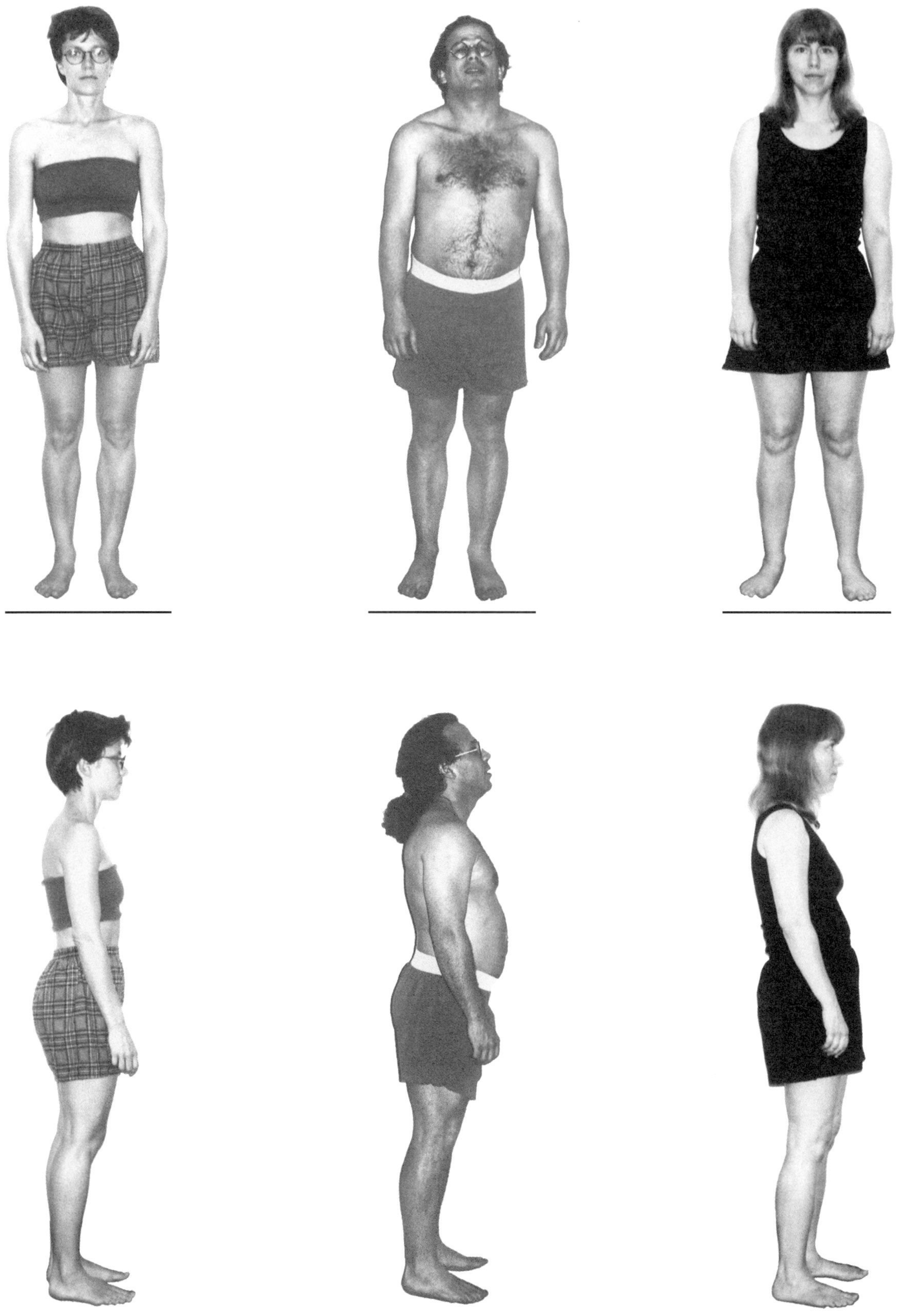

CHAPTER

4

Integrating *Seeing* Into Your Practice

Benefits

People seek treatment from therapists because they want to neutralize the negative effects of trauma, injuries, and accumulated stress from body use and expression patterns. They understand that these accumulated effects will dictate their current amount of ease, strength, and flexibility, and therefore, their long-term quality of life. Many therapists seem to appreciate the ability to use these *seeing* skills to educate their patients about their own body relationships. This enables them to become an educator and not just someone who fixes the isolated problem.

Therapists also say that Aston *Seeing* Skills:

1. Facilitate diagnosis by prioritizing and/or sequencing treatment.
2. Provide a system for refining skills in recognizing abnormal relationships of one body part to another and detecting the causes of many other areas of pain.
3. Offer information about areas of the body that are overutilized or underutilized while performing an activity.
4. Help in applying and integrating other learned techniques.
5. Assist the physical therapist in treatment and in teaching efficient patterns of function movement, which helps prevent the recurrence of pain.
6. Assist in designing a treatment plan that establishes adequate support in the body for a specific task such as an appropriate base of support in standing to facilitate reaching.
7. Are applicable to everyone from the elite athlete to the neurologically challenged patient.
8. Expand the occupational therapist's ability to analyze and identify the client's problems and allow for more effective intervention.

9. Make better use of therapy time and educate patients in optimal use of their bodies while respecting existing limitations.
10. Help in viewing the body holistically, each part exhibiting a cause-and-effect relationship, so you can identify and treat deficits that interfere with the activities of daily living.
11. Help identify which parts of the musculoskeletal system are compressed, overextended, decreased in dimension, hypermobile or hypomobile.
12. Show how the body pattern contributes to fatigue, overuse syndromes, and pain.
13. Show how postural and movement patterns compromise a person's ability to reach, lift, or carry, affecting occupational performance, ergonomics, and leisure activities.
14. Provide a better understanding of the relationships between soft-tissue restrictions and body alignment.
15. Help the personal trainer see where the client needs to increase or decrease tone to improve alignment, stabilize good posture, and increase strength.
16. Help in the design of exercises based on an individual's unique structural pattern, decreasing the risk of injury.
17. Help clients achieve optimal improvement by seeing what the individual pattern was before the exercise pretest and how the exercise subsequently changed that pattern.

In addition, psychotherapists have found that understanding some elements of the relationship of physical or postural expression to a person's internal reality or psyche can offer tools for assisting a client's progress. Alignment and dimension combine to give the body its shape. This shape can reflect or store elements of history, personality, and style of communicating. Psychotherapists often see clients when they are stuck in a loop of experience or behavior that reinforces itself with a corresponding body pattern. The capacity to perceive, and perhaps enable the client to perceive, self-expression in terms of alignment and dimension can be useful in:

1. Providing a tangible model to see and work with the whole person.
2. Identifying and integrating the expression of particular emotional states or self-images in the body (e.g., to differentiate an internal pattern such as fear with an external expression such as sadness or grief).
3. Seeing how a body pattern may be expressing a historical moment that has or has not been worked through.
4. Recognizing how a familiar body pattern may restrict certain new expressions or experiences.
5. Helping the therapist know when to refer a patient for work on physical patterns in order to facilitate the psychotherapeutic process.

Sessions

Following is a suggested sequence to help you prioritize and summarize the information you gather.

Observation Sequence

1. What view gives you the most information?
2. What area gets your attention first?
3. In relationship to? (another area)
4. In relationship to? (the whole body)
5. Consequences on dimension?
6. Possible consequences of tone and mobility on the body
7. Possible challenges from the information above
8. How this information relates to a client's interest (ADLs, sports, pain, etc.)

Sample Session Sequence

This session sequence may be useful in designing your sessions, lessons, and exercises.

The Session Progression

A. Gather Information
    1. Client's focus
    2. History—past injuries, surgeries, structural limitations
    3. Pretest
        For the therapist
            Photo
            action—standing, bending, gait, etc.
        For the client
            self-awareness of posture or function for postcomparison
    4. Notation on body chart.

B. Select and Prioritize Information to Determine the Session Focus and Plan

C. Session/Treatment Plan

   1. Objective
   2. Content:

      modalities

      bodywork

      exercise

      ADLs

   3. Result: Posttest (compare with pretest)

D. Additional Applications to Daily Use: ADLs, Sports, etc.

E. Session Sequence Flowchart

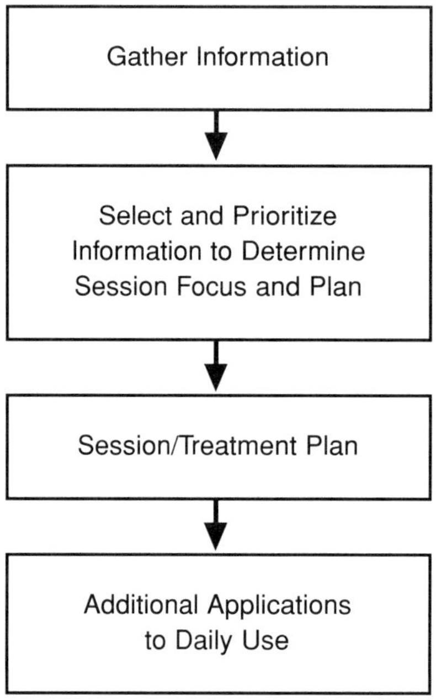

Additional Information for Observing Posture

You can gain even more information by asking your client to stand in different ways. Ask your client to:

1. Stand in his or her best posture. This demonstrates learned postural positions.
2. Stand in his or her casual stance. This demonstrates compensation tendencies.
3. Stand with weight fairly even on both feet. This demonstrates the tendency of the body to be more centered over the left or right foot or over the heels or toes.
4. Stand the same as in No. 3 but with the eyes closed. This demonstrates the change in balance when the client is not using his or her eyes to secure postural position.

Benefits of Photos

I. Benefits of Taking Photos

   A. For therapists

      1. Can record posture in all four views at the beginning of treatment as a reference point.
      2. Can compare changes throughout the client's sessions.
      3. Can document results at the end of treatment.
      4. Can determine the next step in treatment. Can observe certain tendencies in your results that may signify an area that you may not be addressing.

   B. For clients

      1. Encourage their participation in the session.
      2. Photos educate the client. Some people have never seen themselves from the side view or back view. This often is an eye-opening education.

         Some clients prefer *not* to see their photos. In this case, it is better to keep the photos in your file until significant visual changes occur. Then you can have the client observe the difference.

         If the client has a low self-esteem or is highly self-critical, perhaps you could use a transparency to make an abstract body drawing to show him or her.

II. Different Ways to Use Photos

A. Overlay with a transparency to outline the segments and observe alignment and dimension. This creates an abstract body drawing.

B. Use dental floss as the plumb line to easily identify displacements.

C. Overlay with a transparency to notate placement and dimension or to notate placement and dimension on a body chart. Transfer the notations onto a body chart.

D. Turn the photo upside down for a different perspective of how the client would be able to balance if standing on his or her head.

E. Observe one isolated segment (e.g., the head) by covering the rest of the body with a piece of paper. Imagine or draw what you think the position of the other body segments will be. Compare your drawing with the photo.

F. Instead of drawing lines to connect segmental centers on a chart, on a transparency draw the lines to connect segmental centers. Transfer these lines alongside the photo or body chart.

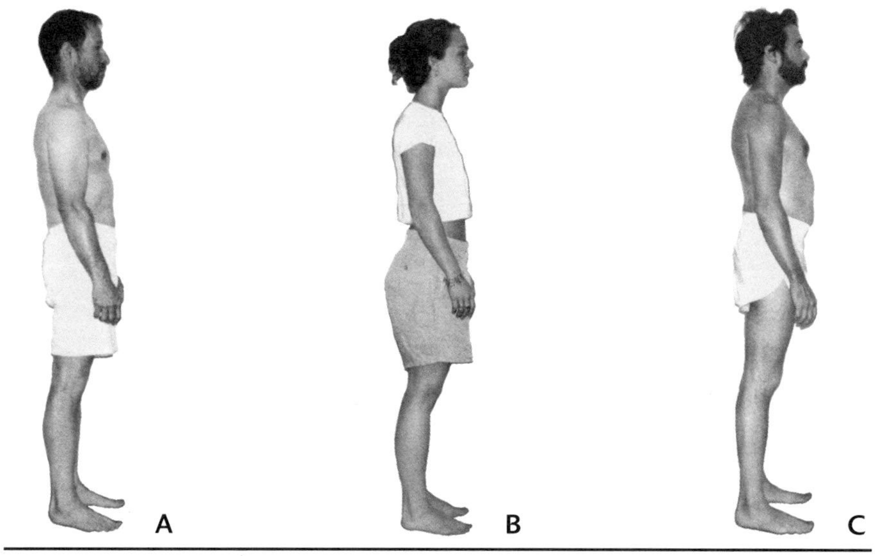

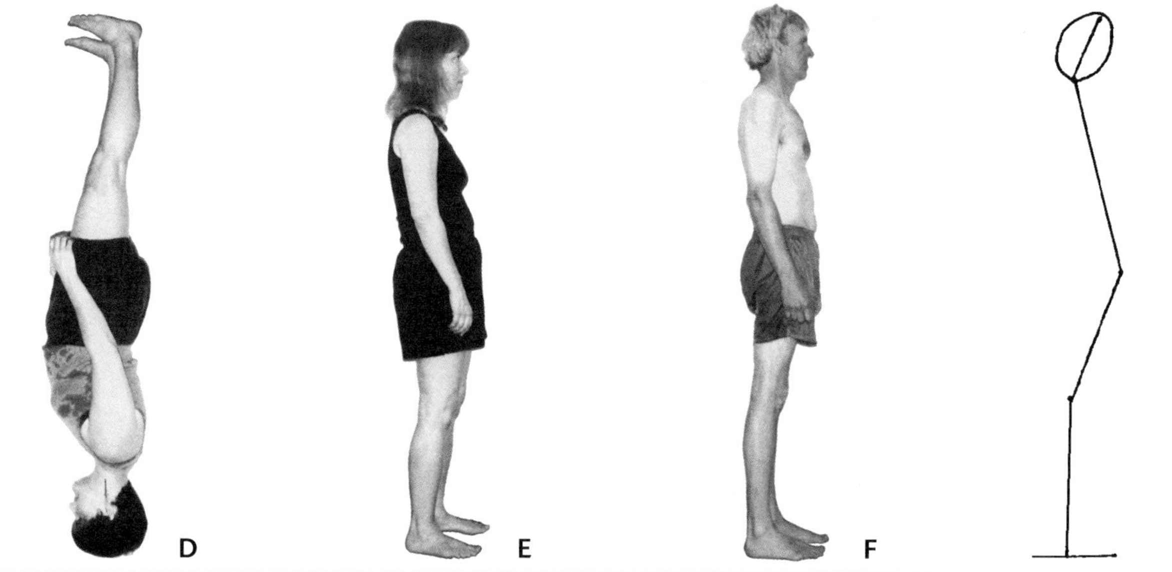

Communicating with Your Patient/Client

How you communicate with your clients will determine not only what they learn but also how they feel about themselves when they hear the information or receive the treatment.

A. Use descriptive words rather than interpretive words (e.g., shorter and longer rather than slouched or depressed).

    1. This enables you to communicate with less emotionally charged words, making the problem more manageable.

    2. This enables clients to be more involved in problem solving with less judgment.

B. Designing Pretests

    1. Include the clients in the process of discovering elements that contribute to discomfort or interest, engaging their participation, curiosity, and questioning.

    2. Depending on the client's self-awareness, have him or her pretest and posttest on a few examples:

        a. the amount of breath he or she can take in a deep inhalation.

        b. where he or she feels the greatest weight bearing on the feet in standing or on impact in gait

        c. the ROM at each specific joint in reaching, side bending, etc.

CHAPTER

5

Applications

This book focuses on seeing body relationships. Although there is not enough space in this manual to go into the content of movement education or ergonomics, perhaps you can appreciate the usefulness of your *seeing* skills in the following examples.

These examples demonstrate familiar body patterns in exercise and everyday activities. Again, I must remind you that every person will need to find his or her best postures and usage for any particular task. Even so, observe the differences in the comparisons of the photos in the following exercises.

Exercise 5.1—Cycling

Observe the differences in the cyclist's body pattern.

Exercise 5.2—Free-Weights

Observe the differences in these two end-range positions of the bench press.

In this photo, observe the arch in his back and his flexed head position.

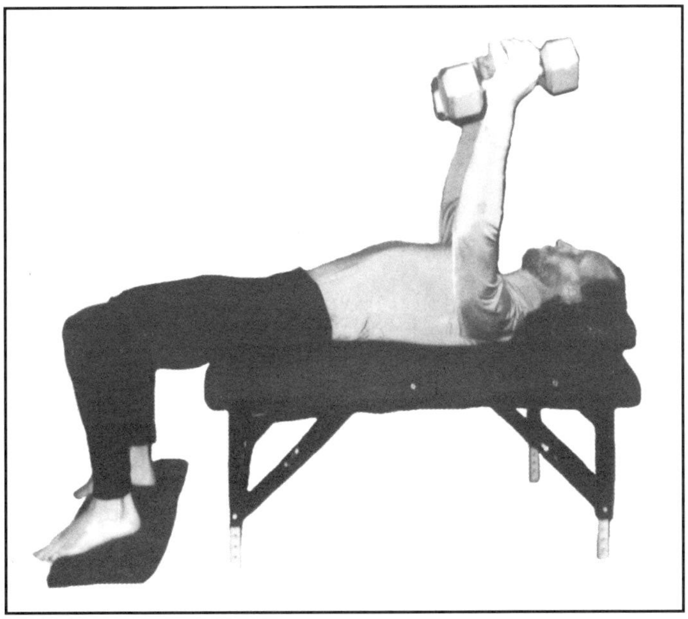

Placing a towel under his head in order to put it into neutral position allows his back to be positioned more safely in neutral as well. The bench is high for his lower leg length, which is remedied by placing a towel or a wedge under his feet.

Exercise 5.3—Backpacking

Observe this hiker carrying 40 pounds of weight in his backpack. The photo on the left demonstrates a familiar body pattern, particularly at the end of a long day of hiking. Compare the body patterns in the two photos.

Exercise 5.4—Daily Activities

Car Seat

There are many variables to negotiate when sitting in the driver's seat of your car—the ergonomics of the car seat, the position and angle of the steering wheel, and the accelerator and brake pedals. You also need to factor in your height and the length of your legs and arms, in order to reach the steering wheel and pedals. You must also consider personal limitations such as low back problems or a stiff neck, which limits your ability to turn your head. Although one could modify the car seat position with wedges, cushions, and lumbar supports (if the car and car seat design allow you to do this), observe the difference in these two photos as a result of changing the sitting posture.

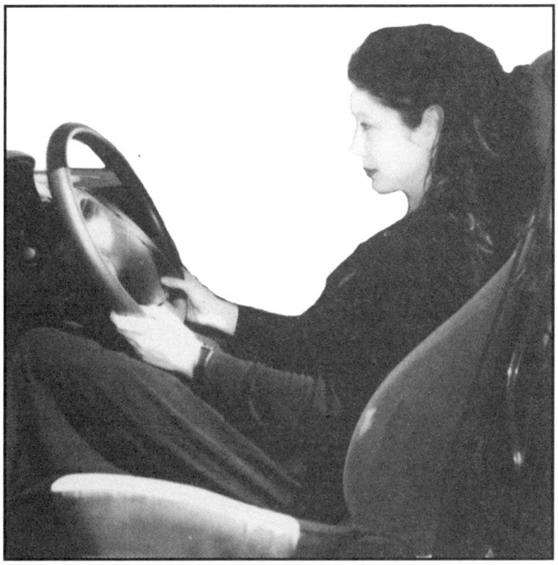

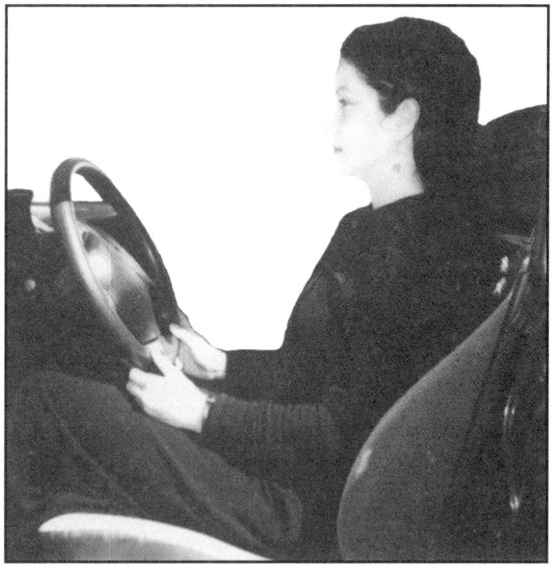

Exercise 5.5—Lying on the Back (Supine Position)

Sleeping or Manual Therapy

When you observe people in the standing position who have good alignment, you can match their standing posture with their body's sleeping position.

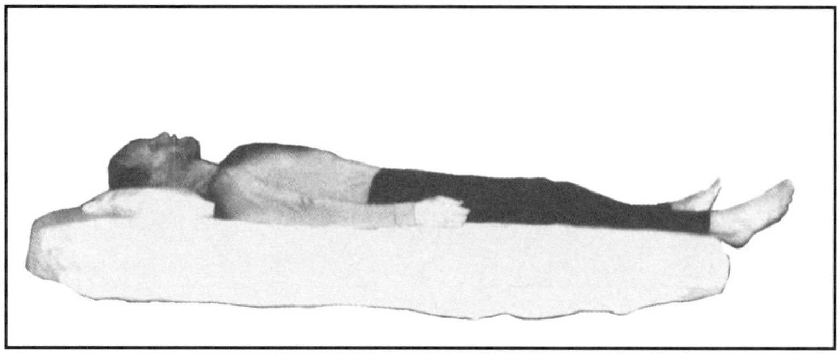

However, if you were working with a client who has osteoarthritis, for example, and whose standing position has to be flexed, you could support this position and, therefore, increase his or her quality of rest by matching the pattern.

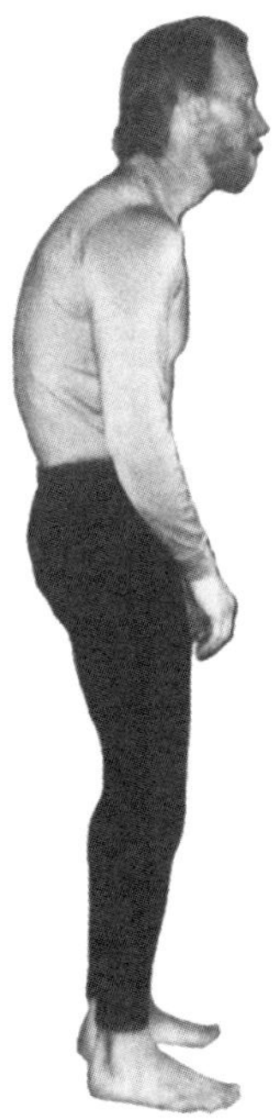

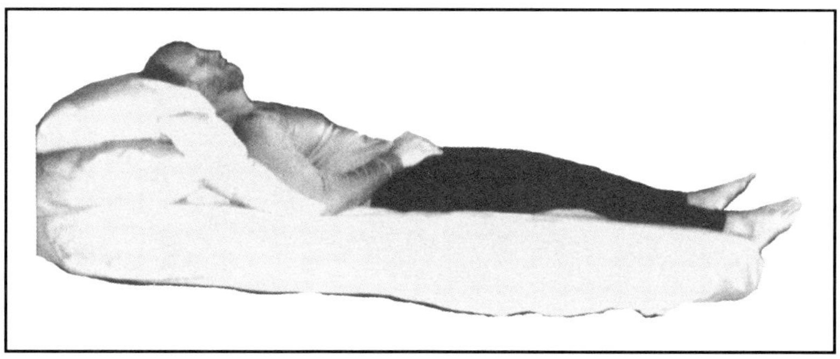

CHAPTER

6

Conclusion

At this time, take a moment to observe three of the figures repeated from the beginning of this book. Use the observation sequence here to guide you.

Observation Sequence

1. What view gives you the most information—the front, back, or sides?
2. What area gets your attention first?
3. This area (from question No. 2) is in relationship to what other area?
4. What relationship does this area have with the rest of the body?
5. What is the consequence on dimension?
6. What is the possible consequence on tone and mobility?
7. What are the possible challenges for the body based on the above information?
8. How might this information relate to the client's concerns about his or her symptoms?

Exercise 6.1

Looking at the following figures, can you now identify why the top segments (upper chest, neck, and head) do not belong to the lower body patterns? Before you look at the next page, see if you can give three reasons why they do not match. Now look at the next page and compare these figures with the before-sessions photos.

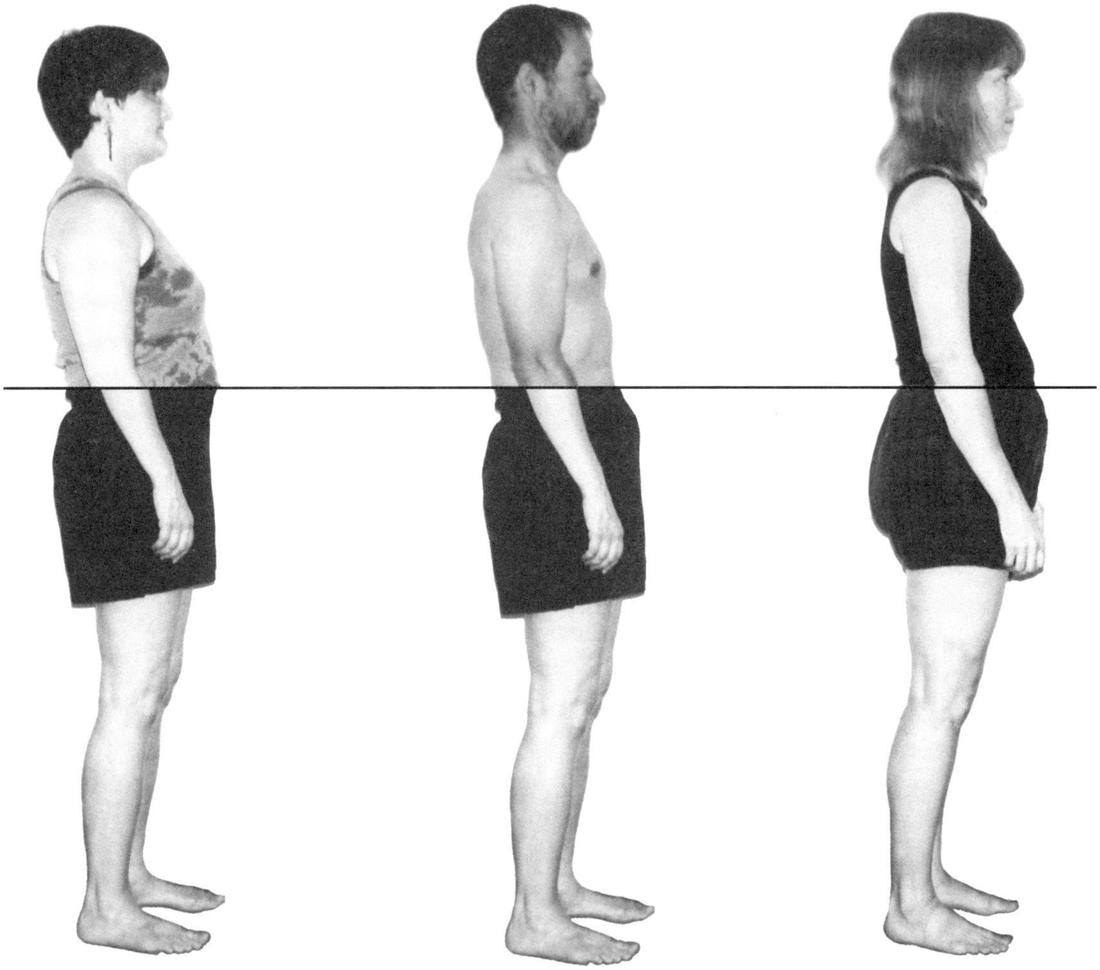

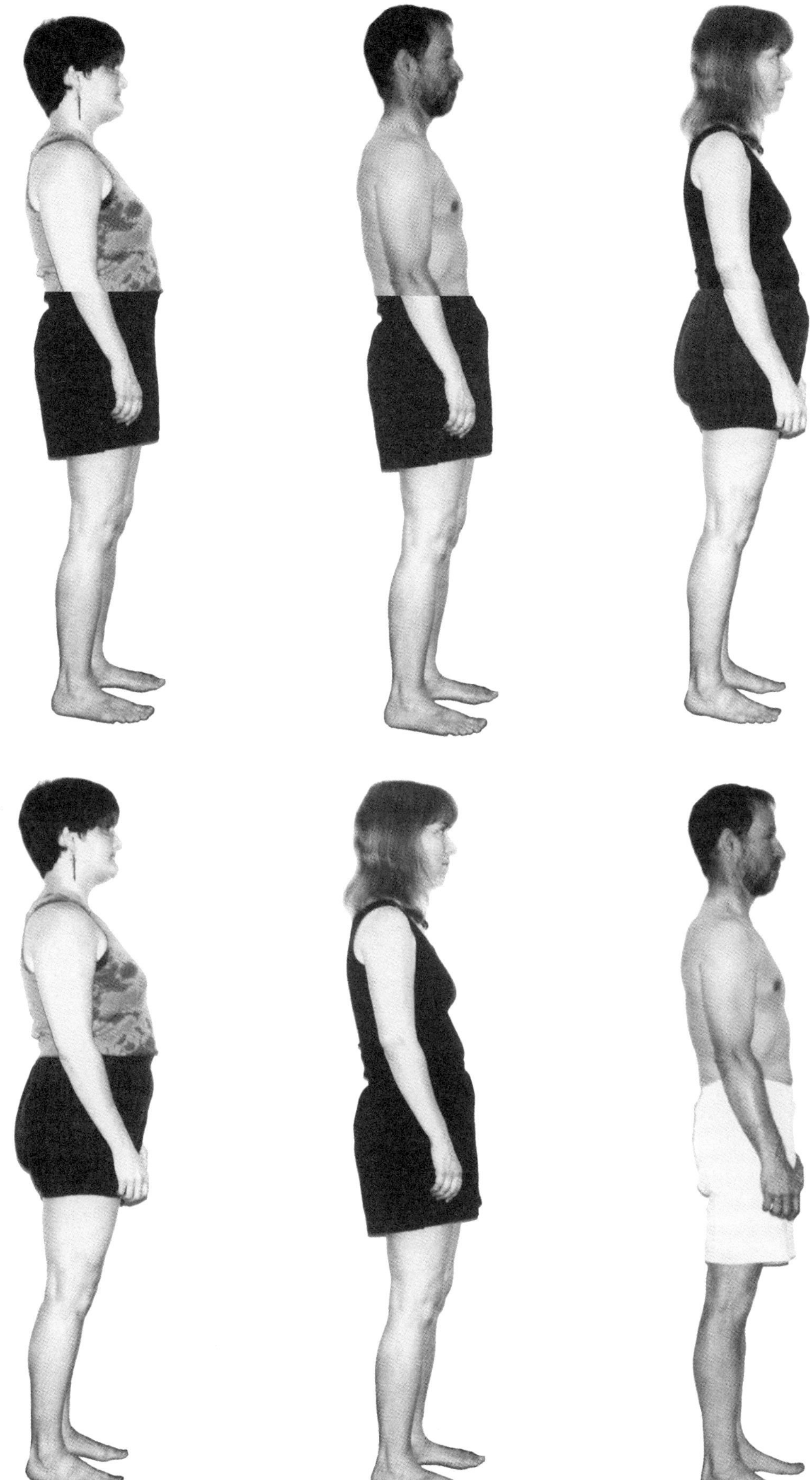

Posttest

Now that you have completed this book, do the following.

1. Reassess your pretest photo. Jot down your observations or use a transparency to record your assessment.
2. Compare this assessment with your pretest assessment.
   a. How do your present observations compare with your pretest observations?
   b. Does this information change your plan of treatment or lesson?
3. Now compare the photos from the client's before-and-after session.

The left side

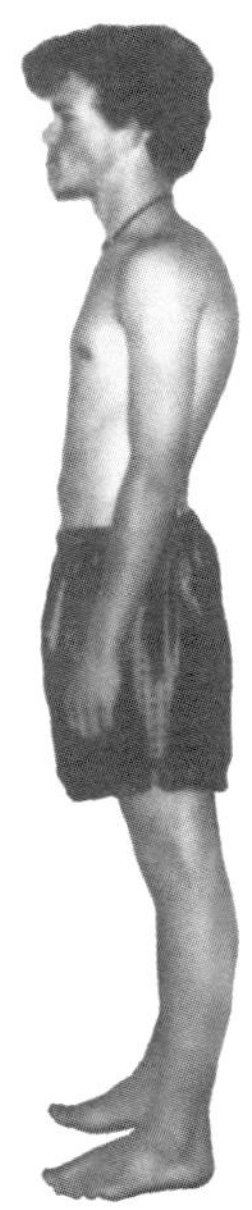

June 2

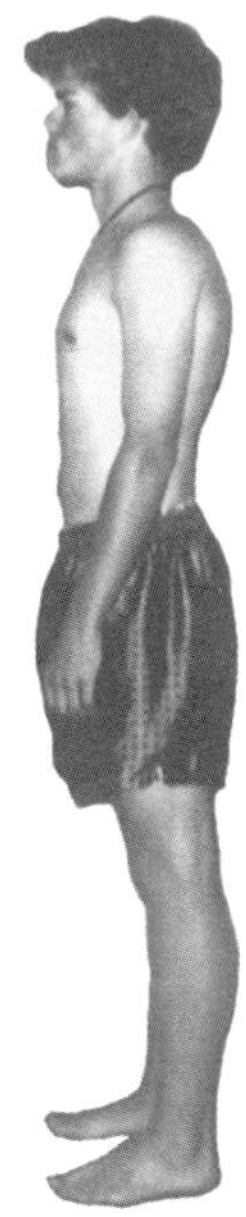

June 2
(after his first session)

The right side

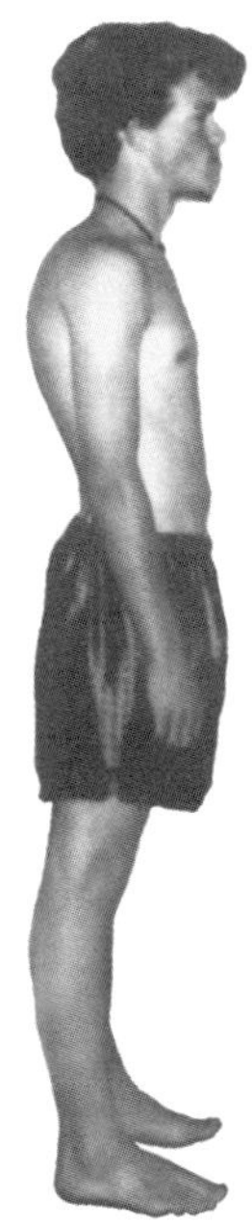

June 2

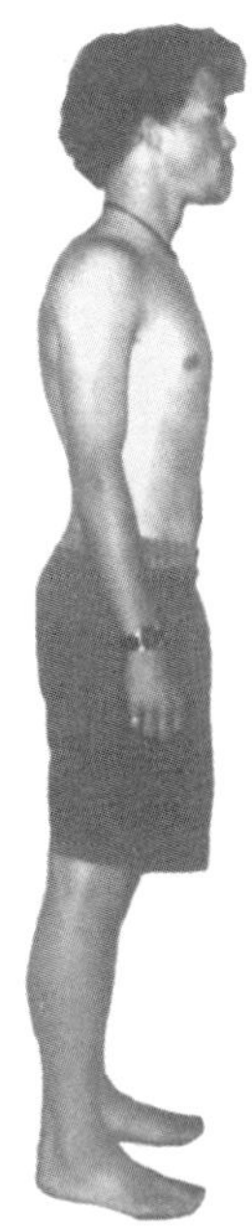

June 8
(after his second session)

Testing Your *Seeing* Skills

Compare these before- and after-sessions photos. This time, start your assessment from seeing dimensional changes. Then relate them to alignment and possible hypertone and hypotone mobility patterns.

What changes transpired as a result of the client's six sessions?

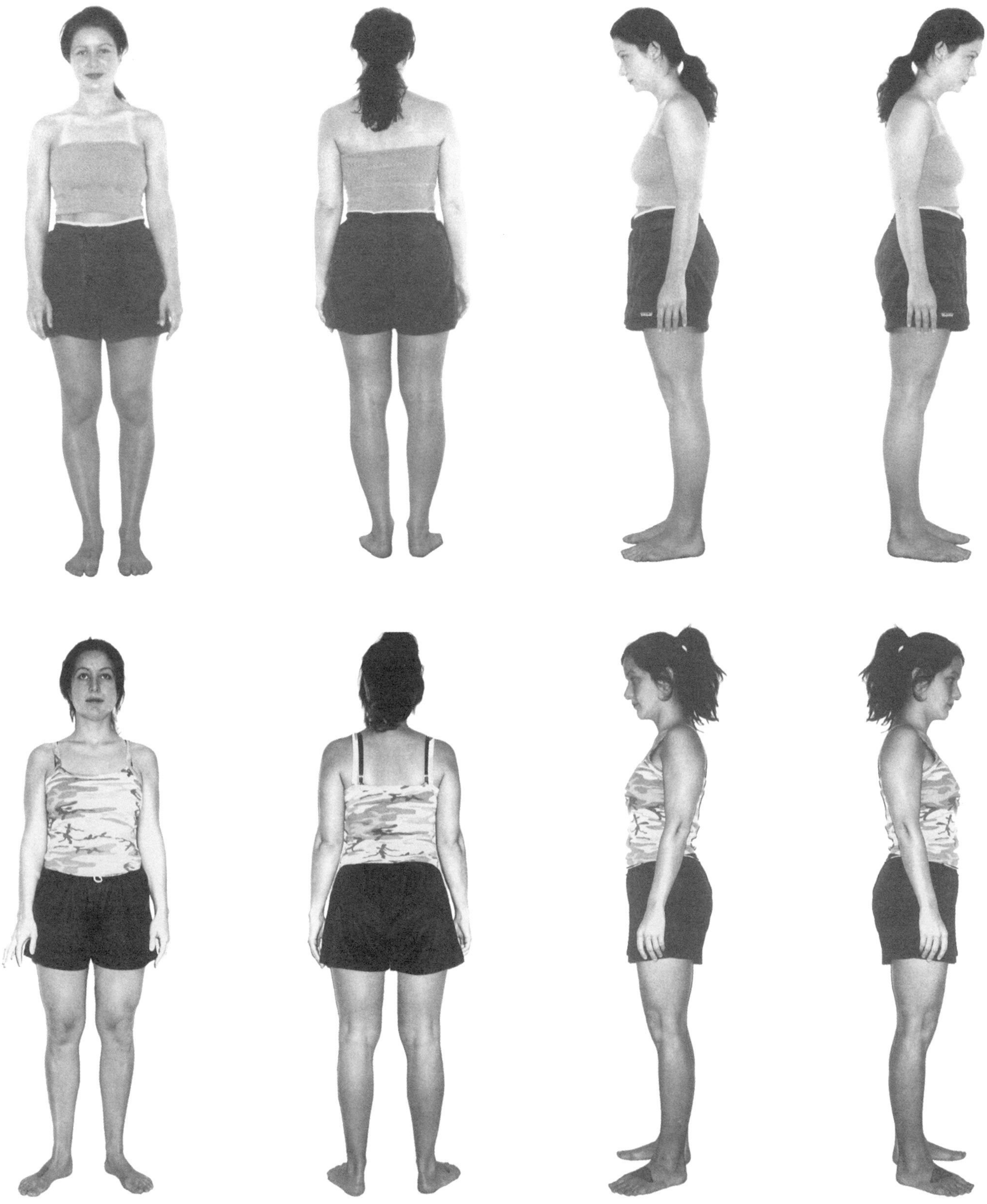

Afterword

We have covered quite a bit of information for you to assimilate. Many of you already may have acquired great skill at seeing the body alignment in relationship to the plumb line. Perhaps you can focus more on the available information by observing the dimensional aspects, the forces of gravity and GRF, the Aston-Mechanics guidelines for asymmetry, range of neutral (RON), and alignment and function. I hope you will use whatever portion of this book best meets your interests and needs.

The first six drafts of this book had enough content for at least three books. Although it was a challenge to have to eliminate information on movement education, ergonomics, and fitness, space limitations precluded doing justice to these subjects.

I trust that you know by now that this information is to be used as a guideline, as reference to your current understanding and as an invitation for future exploration of the many applications of these ideas about posture, movement, exercise, ergonomics, expression, and about just living!

Please remember that what appears to be a model is used as a point of departure for study; in actuality, every body/mind has its current best neutral position and its own emerging model.

–Judith Aston

Appendix

1. Observation Sequence
2. Sample Session Sequence
3. Sample Session Form

Observation Sequence

1. What view gives you the most information—the front, back, or sides?

2. What area gets your attention first?

3. This area (from question No. 2) is in relationship to what other area?

4. What relationship does this area have with the rest of the body?

5. What is the consequence on dimension?

6. What is the possible consequence on tone and mobility?

7. What are the possible challenges for the body based on the above information?

8. How might this information relate to the client's concerns about his or her symptoms?

Copyright © 1998 by Therapy Skill Builders, a division of The Psychological Corporation. All rights reserved. Aston Postural Assessment Workbook: Skills for Observing and Evaluating Body Patterns / ISBN 076161530X / 1-800-228-0752 / This page is reproducible.

Sample Session Sequence

This session sequence may be useful in designing your sessions, lessons, and exercises.

The Session Progression

A. Gather Information
   1. Client's focus
   2. History—past injuries, surgeries, structural limitations
   3. Pretest

      For the therapist

         Photo

         action—standing, bending, gait, etc.

      For the client

         self-awareness of posture or function for postcomparison
   4. Notation on body chart.

B. Select and Prioritize Information to Determine the Session Focus and Plan

C. Session/Treatment Plan
   1. Objective
   2. Content:

         modalities

         bodywork

         exercise

         ADLs
   3. Result: Posttest (compare with pretest)

D. Additional Applications to Daily Use: ADLs, Sports, etc.

E. Session Sequence Flowchart

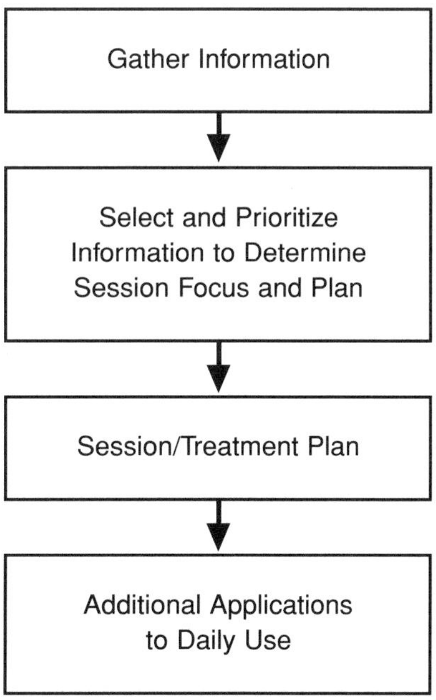

opyright © 1998 by Therapy Skill Builders, a division of The Psychological Corporation. All rights reserved. Aston Postural Assessment
orkbook: Skills for Observing and Evaluating Body Patterns / ISBN 076161530X / 1-800-228-0752 / This page is reproducible.

Sample Session Form

Client's Name:____ Date:____________________

Subjective (Client's focus):

Circle area(s) of complaint.

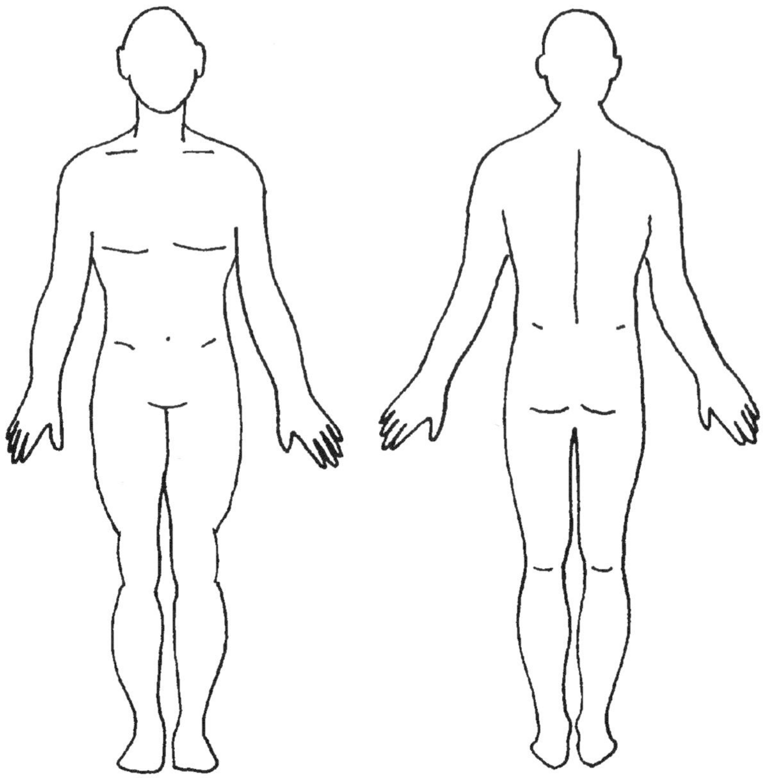

Objective (Your observation of alignment, dimension, etc.):

Assessment (Hypothesis):

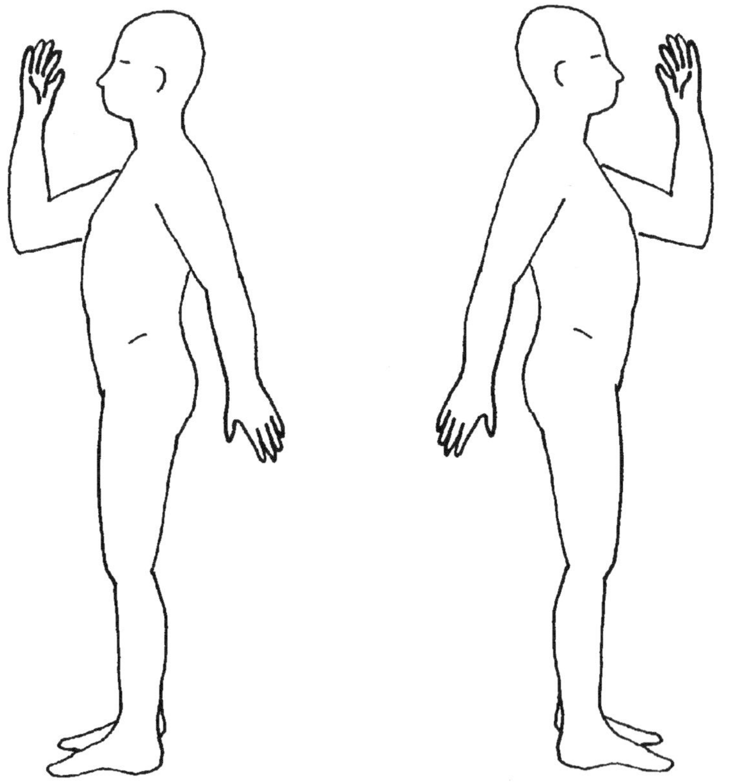

Plan (Session sequence):

Copyright © 1998 by Therapy Skill Builders, a division of The Psychological Corporation. All rights reserved. Aston Postural Assessmen
Workbook: Skills for Observing and Evaluating Body Patterns / ISBN 076161530X / 1-800-228-0752 / This page is reproducible.

Glossary

acetabulum—The socket or hollow cup in the pelvic bone into which the femoral head fits. These bones comprise the hip joint.

aperture—A term used in Aston-Mechanics to refer to one aspect of the cross-sectional areas of the body. An example of an aperture would be any two-dimensional cross-section of a body segment. These intersections of segments behave like the iris of the eye or the aperture of a camera, in that they increase, decrease, or change shape depending on their relationship to the adjacent segments.

anteversion of the hip—an abnormal position of the femur in which there is forward displacement.

ASIS (Anterior/Superior Iliac Spine)—a bony prominence on the anterior aspect of the iliac crest (pelvic bone).

biomechanics—the scientific study of how internal and external forces act on the living body.

calcaneus—commonly known as the heel bone.

clavicle—commonly known as the collar bone, it attaches to the sternum (breast bone) and to the acromium process of the scapula (located near the shoulder joint).

coccyx—the bottom distal portion of the spine, generally known as the tailbone.

costal arch—the distal border of the rib cage on the front of the chest.

ergonomics—the science of equipment and product design to promote good posture, efficient movement, and reduced fatigue. The study of efficient movement and good posture.

femur—the thigh bone, which is connected proximally to the hip joint, and distally connected to the knee joint.

flexion—the process of decreasing an angle.

greater trochanters—bony prominences located at the proximal portion of the femur, to which muscles are attached.

humerus—the upper arm bone, which is attached proximally to the shoulder joint and distally attached to the elbow joint.

hypermobility—a condition in which a joint has excessive range of motion.

hypomobility—a condition in which a joint has a limited or less than average range of motion.

hypotonicity—limp or below average muscle tone while at rest, with movement, or both.

hypertonicity—an increase in muscle tone, either at rest or with movement, or both.

ilium—the largest part of the pelvic bone, consisting of the upper portion of the os coxa.

iliac crest—the superior part of the pelvic bone, which spans from the ASIS (anterior/superior iliac spine) to the PSIS (posterior/superior iliac spine).

ischial tuberosities (IT)—the bilateral, bony prominences of the ischium. These prominences bear the body's weight when a person is sitting.

ischium—the distal part of the os coxa (pelvic bone).

lordosis—an increase in the natural curve of the lumbar spine.

malleoli—the distal bony prominences located on either side of the ankle joint. The plural form of malleolus.

palpation—the use of the hands to feel soft tissue, muscle, or bone.

PSIS (Posterior/Superior Iliac Spine)—a bony prominence on the posterior aspect of the iliac crest (hip joint).

range of neutral—the body's continuous movement or sway through its changing central axis.

sagittal plane—an abstract, imaginary plane that divides the body into the left and right sides.

scapula—commonly referred to as the shoulder blade. It is made up of many bony prominences, and it articulates with other bones to provide joints that promote shoulder flexibility.

scoliosis—abnormal, lateral spinal curves, usually in the thoracic and lumbar regions, causing the spine to deviate from a straight line.

sternal notch—the narrow space where the two clavicles meet at the top of the sternum.

tensile forces—two equal and externally applied forces acting along the same line but in opposite directions.

Temporal Mandibular Joint (TMJ)—the joint located just anterior to the ear that connects the mandible (lower jaw) to the temporal bone of the skull.

trapezius muscle—a large, superficial muscle of the back that is shaped like a triangle. Its bony attachments are from the base of the skull and the cervical and thoracic vertebrae to the clavicle and scapula.

viscera—the organs located within the body cavity.